Separation Process Essentials

Separation Process Essentials

Alan M. Lane

CRC Press
Taylor & Francis Group
Boca Raton London New York

CRC Press is an imprint of the
Taylor & Francis Group, an **Informa** business

CRC Press
Taylor & Francis Group
6000 Broken Sound Parkway NW, Suite 300
Boca Raton, FL 33487-2742

International Standard Book Number-13: 978-1-138-08608-1 (Hardback)

Library of Congress Cataloging-in-Publication Data

Names: Lane, Alan M., author.
Title: Separation process essentials / by Alan M. Lane.
Description: First edition. | Boca Raton, FL : CRC Press/Taylor & Francis Group, [2020] | Includes bibliographical references and index.
Identifiers: LCCN 2019030705 | ISBN 9781138086081 (hardback ; acid-free paper) | ISBN 9781315111131 (ebook)
Subjects: LCSH: Separation (Technology) | Extraction (Chemistry)
Classification: LCC TP156.S45 L36 2020 | DDC 660/.2842—dc23
LC record available at https://lccn.loc.gov/2019030705

Visit the Taylor & Francis Web site at
http://www.taylorandfrancis.com

and the CRC Press Web site at
http://www.crcpress.com

Contents

Part III Absorption and Stripping

Part IV   Solvent Extraction

Part V   Membranes

Preface

Many good separation process textbooks are available for chemical engineering. So why write yet another one? Mainly, I just wanted to write a textbook that reflected the way I taught the separations course at The University of Alabama for the past 30 years. Students left my course with a solid understanding of separation processes without being overwhelmed by details and impossible to memorize derivations and correlations. I approached CRC Press in 2016 about publishing my book. They conducted a survey of professors that teach this subject. The majority were open to considering a new book and perhaps you will consider mine. As you will see, my book introduces just the essentials of separation processes.

Most engineering textbooks attempt to be comprehensive and cover way more than can be learned in a one-semester introductory course. The advantage is that they can also be used for advanced and graduate-level courses and as a reference for practicing engineers. The disadvantage is that they become dense, filled with pages of mind-numbing derivations, and difficult to sort out the essential information. Let's be honest: Most students don't actually read them. Mostly, they just look at the assigned homework problems and then search the chapter for the right equation or example to follow.

I hope you find my book a bit unconventional. I intend for you to actually read it and work through the examples with pencils in hand and spreadsheets open, almost like a workbook. I wrote it while imagining that you were sitting right across the table from me as I explain the material. Go ahead and "deface" the book by writing in it – I left plenty of blank spaces. Solve the problems using the many Excel spreadsheets that can be downloaded from my website (SeparationsBook.com). Use these tools to dream up every possible scenario that your professor could put in an exam.

Only the most common processes will be discussed: distillation, absorption, stripping, and solvent extraction. These can all be treated as equilibrium stages. This means that two phases are contacted, the chemicals distribute between the phases until they reach equilibrium compositions, then they are separated. Usually these stages are stacked together in towers for even better separation and the two phases flow countercurrent. No doubt you've seen the tall distillation towers as you drive past any chemical plant or refinery. Membrane separations are thrown in at the end as an example of a non-equilibrium process.

Interestingly, these processes are all described by just a few equations that *you already know* – mole balances, phase equilibrium relationships (like Raoult's law), various process specifications, and, sometimes, energy balances. There is very little here that you haven't already mastered in the introductory chemical engineering course in mass and energy balances.

Each process is developed in the same way. First, a single equilibrium stage is analyzed using the above equations. A flash drum, for instance, is a single-stage distillation process. Then we see that stacking several stages produces even better separation. Often, many tens of stages will be present in a distillation column. Finally, plotting the compositions of the internal stream compositions leads to graphical techniques that can simplify the analysis. This only answers one question, but it is the big one: "How many equilibrium stages are required to accomplish a specified separation?" Other important questions like "how many real stages are required," or "how tall and wide a column should be," will be briefly addressed near the end.

Well, that's all I want to say in this preface other than where to find supplementary material. You will find lots of free downloads at SeparationsBook. com. This is where you will find all of the Excel spreadsheets referenced in the text as Excel "name/page." Most problems require solving multiple, coupled algebraic equations and these spreadsheets are designed to be user-friendly, with detailed instructions, color-coded cells, and suggestions for modifications. If you ever hopelessly mess up a spreadsheet, you just need to go back and download the original again. Another section contains all of the chapter problems. This is intended to be a "living collection" that is continually expanded. Everyone, students and teachers alike, is welcome to email new and interesting problems to me and I'll add them to the party. Anything else that I can think of to help you learn this material will be on this page, including YouTube tutorials, discussion boards, links to classic papers, learning tools that I come across, and a list of mistakes (errata) that are likely to crop up in these early editions. Please email me with suggestions at alane@eng.ua.edu.

Happy learning!
Dr. Alan M. Lane
Professor Emeritus of Chemical and Biological Engineering
Center for Materials for Information Technology
The University of Alabama
Tuscaloosa, AL 35487-0209

Author

Dr. Alan M. Lane is Professor Emeritus of Chemical and Biological Engineering at The University of Alabama. He has worked for Union Carbide Corporation (polymerization processes, 1984–1986), Battelle's Pacific Northwest Laboratories (nuclear waste treatment, 1977–1979) and Pacific Northwest Testing Laboratories (ASTM testing, summers 1968–1976). He has been a visiting scholar at Boise Cascade Corp. (dioxin from pulp bleaching, 1990), Qingdao Institute of Chemical Technology (chemical reactor modeling, 1993), the University of Wales (magnetic ink characterization, 1995), and Argonne National Laboratory (fuel processing to make hydrogen, 1999). Lane earned BS degrees in both Chemistry and Chemical Engineering from the University of Washington (Seattle) in 1977. He obtained a PhD degree in Chemical Engineering from the University of Massachusetts (Amherst) in 1984. His academic research projects have covered a broad spectrum of chemical reaction engineering, especially heterogeneous catalysis: chemical reactions during metal casting, dioxin formation during waste incineration, selective synthesis gas reactions, green manufacturing, diffusion in porous media, hydrogen production for fuel cells, and fuel cell electrode reactions. He also studied the complex rheology of magnetic inks for magnetic tape and synthesis of rare-earth-free permanent magnets. His teaching interests are focused in chemical reaction engineering, unit operations laboratory, and separation processes. Professor Lane is also a singer-songwriter performing in dives across the south as Dr. Doobie 'Doghouse' Wilson (DoobieDoghouseWilson.com). His five albums of original music include chemical engineering-themed songs like the #1 "Stepping Off The Stages" about distillation column design (Lane, 2008).

Part I

Introduction

The book is divided into several parts. This part needs no introduction. It is the introduction!

But as long as you are here, Chapter 1 "Introduction" will show you how I think about separation processes, my teaching philosophy and methods, important features to look for, and how the book is laid out. A bit of that is in the Preface too. The Preface was never assigned when I was an undergraduate and I rarely read it. I hope you go back and read those few pages.

Chapter 2 "A Look Inside Your Chemical Engineering Toolbox" is an essential review of the skills you learned in previous courses, especially the first chemical engineering course in material and energy balances. You already possess almost every tool you need for this subject. We are simply going to apply them to the specific process of separations.

1

Introduction to Separation Processes

So, here you are, in the third year of your chemical engineering studies and ready for a course on separation processes. Congratulations! That's awesome! Questions?

Q: Why should I learn from your book?
A: There are plenty of other good books on separation processes. Some are listed in Table 1.1 with the first four being the most popular in the United States. Why use this new book *Separation Process Essentials* by Professor Emeritus Alan M. Lane?

My book is just for you, a chemical engineering student being introduced to separations for the very first time. I do not cover all separation processes, just the most common ones. So, it won't be a great reference book for specialists. I limit the content to what you can reasonably learn in one semester. So, it won't be appropriate for advanced or graduate studies. Most chemical engineering students no longer study computer programming, so I limit the computer tools you need to spreadsheets and (optionally) CAD software. I am a big believer in learning with examples, and you will find lots of thoroughly explained exercises throughout. As I write this, I try to imagine you sitting in my office and informally discussing the topic. I hope you find it easy to read, even humorous in parts. After 30 years of teaching this stuff, I have a pretty good feel for what you can learn well in one semester.

When and if you need a broader or deeper knowledge of separation processes, you can always take an advanced or graduate course, read a more comprehensive book, work with your company's design group, and talk to vendors of separation equipment.

TABLE 1.1

Some Textbooks on Separation Processes

1. Separation Process Engineering (4th Edition) by Phillip C. Wankat (2016)
2. Separation Process Principles (4th Edition) by J. D. Seader, Ernest J. Henley, and D. Keith Roper (2015)
3. Transport Processes and Separation Process Principles (5th Edition) by Christie John Geankoplis, A. Allen Hersel, and Daniel H. Lepek (2018)
4. Unit Operations of Chemical Engineering (7th Edition) by Warren L. McCabe, Julian C. Smith, and Peter Harriott (2004)
5. Separation Processes (2nd Edition) by C. Judson King (2013)
6. Mass Transfer and Separation Processes (2nd Edition) by Diran Basmadjian (2007)
7. Principles and Modern Applications of Mass Transfer Operations (3rd Edition) by Jaime Benitez (2016)

Q: What do you mean by "separation processes?"
A: I think you already have a pretty decent idea of what separation processes are. You've passed by (or maybe even worked in) chemical plants with those giant distillation towers you can see from the road. You studied partial evaporation – flash distillation – in the material and energy balance and thermodynamics courses. You even have run across separation processes in your everyday living, like brewing a tasty cup of coffee!

Separation processes are important and ubiquitous in the chemical industry. Most chemical processes are centered on a reactor or some device that converts raw materials into more valuable products. Before the reactor, separators are used to purify the raw materials. After the reactor, separators are used to recycle raw materials back into the process, purify the products, and remove environmental pollutants.

You probably know that distillation is the most common separation process. It is used for materials with reasonable differences in volatility. But many other types exist, as shown in the brief listing in Table 1.2. The choice depends on the physical and chemical differences that you can exploit between the materials to be separated.

Q: How do separations fit into the chemical engineering curriculum?
A: I'll explain with the chemical engineering building blocks shown in Figure 1.1. In the first year, you learn the science fundamentals that form the foundation for chemical engineering. These include chemistry, physics, biology, and math. In the second year, you become skilled at material and energy balances, thermodynamics, and transport phenomena. Material and energy balances allow you to count what goes into a process and what comes out (or accumulates!). Sounds simple but can get quite challenging, as you discovered! Thermodynamics shows how far a process can go, and transport shows how fast it can get there. In the third year, you apply all these skills toward the analysis and design of the major unit operations:

TABLE 1.2

Examples of Separation Processes

Distillation	Heat is added to a liquid (usually) creating a vapor rich in the more volatile compounds and a liquid rich in the less-volatile compounds
Stripping	A gas is added to a liquid causing the more volatile or less soluble components to leave with the gas
Absorption	A liquid is added to a gas causing the less-volatile components or more soluble components to leave with the liquid
Solvent extraction	An immiscible liquid is added to a liquid causing the components to separate into the two liquid phases
Crystallization	Heat is removed from a liquid causing some components to crystallize into solids
Membranes	A liquid or gas is passed over a thin film. Components that are soluble and diffuse faster concentrate on the other side of the film

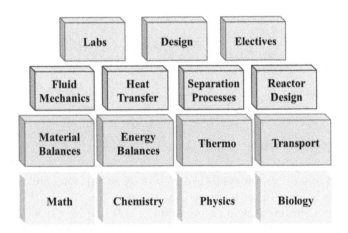

FIGURE 1.1
Chemical engineering building blocks.

fluid flow, heat transfer, chemical reactions, and the topic at hand, separations. Finally, in the fourth year you get to practice chemical engineering in the lab and design courses, including design and operation of separation equipment.

Q: What is the basic idea behind separation processes?
A: Most separation processes consist of mixing two separable phases like gas with liquid; gas or liquid with solid; or even two immiscible liquids. The contact area between phases should be large enough and maintained long enough for transfer of chemicals to occur between them. The distribution of the chemicals between phases approaches thermodynamic phase equilibrium. Finally, the two phases are physically separated.

Think of a simple tank containing a liquid mixture of chemicals. Add heat to partially evaporate the liquid and the more volatile chemicals will concentrate in the vapor (distillation). Bubble a gas through the liquid and some chemicals will pass from the liquid to the gas (stripping) or from the gas to the liquid (absorption). Add a second immiscible liquid and the chemicals will distribute according to their relative solubility in the two liquid phases (solvent extraction).

Maybe you had a delicious cup of hot coffee this morning? You (or maybe a barista) added hot water to ground up coffee beans and some of the chemicals dissolved from the beans into the water (extraction). Then the solids and liquids were separated by passing the mixture through a paper filter (filtration). All sorts of separation processes occur after that in your digestive system – but we won't get into that!

Think about the process variables that affect coffee brewing. How do temperature, mixing, contact time, particle size, amounts, etc. affect the extraction? What are the underlying physical reasons? Close the book and

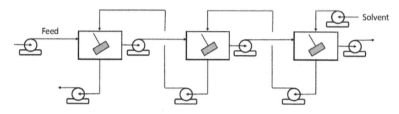

FIGURE 1.2
A series of equilibrium stages.

think about that for a moment. Make a list of these variables and their effects. Discuss this in class or with your study group.

Now take a similar look at carbonation of soft drinks. Consider a tank of flavored water through which we pass bubbles of carbon dioxide. What are the relevant process variables? How do they affect the amount of dissolved carbon dioxide? Make another list of process variables and their effects.

If that single tank of liquid hasn't given you enough transfer of chemicals, you can always add the liquid product to another tank and repeat the process. Perhaps, it simply gives the phases more time to reach thermodynamic equilibrium or maybe the second tank is under different process conditions that enhance the rate of transfer or the equilibrium concentrations. Usually, the most efficient arrangement is to run the two phases through a series of tanks counter currently as shown in Figure 1.2. Each tank is considered an equilibrium stage, although in practice the separation may not reach complete thermodynamic phase equilibrium.

All those tanks, pumps, piping, etc. take up a lot of real estate and can be very expensive. Instead, most separations take place in vertical columns divided into sections that represent the series of tanks as shown in Figure 1.3. The liquid phase goes in the top and leaves at the bottom, while the gas phase goes in the bottom and leaves at the top. The flow is determined by density differences, even when both phases are liquids. In each section, the phases are intimately mixed, held for a short time, and then allowed to separate.

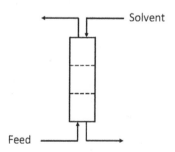

FIGURE 1.3
Simplified schematic diagram of a separation column showing three equilibrium stages.

Q: So, what exactly are these equilibrium stages?

A: You can't see this from the outside (unless it is a glass column like the one in The University of Alabama's unit operations laboratory shown in Figure 1.4), but inside the column are structures – either trays or packing – that promote liquid and vapor mixing and then separation. Each tray and the space above it is an equilibrium stage in which the liquid and vapor compositions reach thermodynamic phase equilibrium – well, in theory anyway, if they are 100% efficient. Packing does not have discrete sections, so we consider a certain height of packing as being equivalent to an equilibrium stage.

We will analyze each stage using engineering tools that *you already know*. Material balances state that the flow of a chemical coming into a stage is the same as that flowing out of a stage (unless accumulation or reaction takes place). Energy balances state that the energy coming into a stage as enthalpy with the entering streams plus the energy coming in by heat transfer is equal to the energy leaving a stage by the same mechanisms. Phase equilibrium equations describe the compositions of the two phases in contact. At this point, we can perform a degree of freedom analysis to see if there are equal numbers of equations and unknowns. If not, we can make a variety of process specifications.

Q: Can you be a little more specific about what these look like?

A: Let's visualize the distillation column before we begin to analyze the process. Similar designs are used for absorption and stripping processes and even for solvent extraction. A more quantitative discussion will be presented in Chapter 16 "Column Design."

You have probably seen distillation columns when driving by any chemical plant or petroleum refinery. You notice tall (50–100+ ft) cylindrical structures

FIGURE 1.4

The University of Alabama's glass distillation tower. The bottom of the column (left) includes the thermosiphon reboiler. The top of the column (right) includes the total condenser.

(Figure 1.5) and if you look closely there is usually a rather large diameter pipe (with low-density vapor) coming off the top. Less obvious are smaller diameter pipes (with high-density liquid) entering at the middle and bottom of the column. If you could look even more closely, you would see that some of these pipes go through large heat exchangers.

Figure 1.6a is a simple process flow diagram (PFD) of the most common distillation process. The most efficient location for the feed is often in the middle of the column. Vapor leaves the top of the column and enters a total condenser that changes the vapor to a (we'll assume) bubble point (saturated) liquid. That liquid is divided into a "light" product, distillate, and a recycle stream, reflux. The condenser is usually placed on the ground because installation and maintenance are difficult for people hanging from harnesses 100 ft in the air! However, you will often see it drawn as if it were hanging precariously from the top as shown in Figure 1.6b. A vapor product could be taken, but liquids are easier to handle, being cooler and ~1,000 times denser. The reflux is pumped back up to the top of the column. The liquid leaving the bottom of the column is partially evaporated in a partial reboiler. The vapor, boilup, is returned to the bottom of the column and the remaining liquid, bottoms, is the "heavy" product. Light and heavy refer to the volatility of the components being separated.

Inside the column are structures – either trays or packing – that promote liquid and vapor mixing and then separation. Figure 1.7a is a drawing of a column section with trays. The liquid flows through a downcomer from Tray 1 to Tray 2, then across Tray 2 to a weir that maintains a certain depth of

FIGURE 1.5
Commercial distillation columns. Magnitrol International UK, "Level Measurement Solutions for Distillation Columns", Process Industry Reformer, May 3, 2018. Reprinted with permission from ProcessIndustryInformer.com.

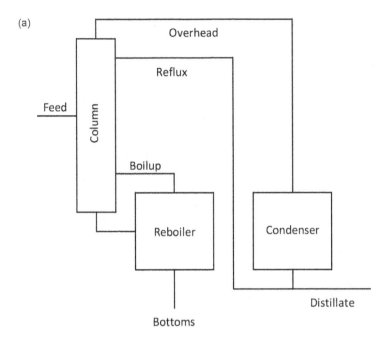

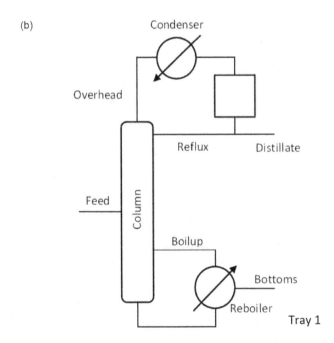

FIGURE 1.6
(a) Simple PFD of a distillation column and associated equipment and (b) distillation column as typically drawn.

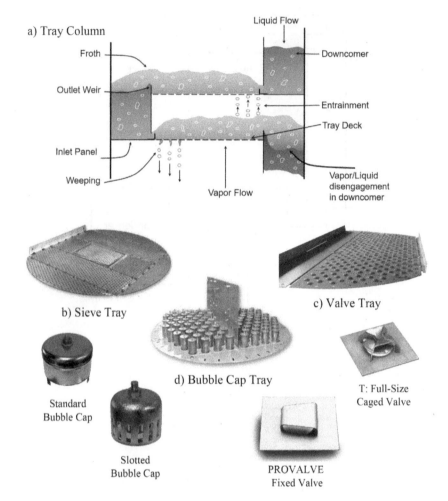

FIGURE 1.7
(a) Drawing of distillation column section with sieve trays. Types of trays include (b) sieve, (c) fixed valve (with two examples), and (d) bubble cap (with two examples). Images of trays and valves courtesy of Koch-Glitsch, LP, Wichita, KS, USA. Images of bubble caps courtesy of Sultzer Chemtech Ltd., Switzerland.

liquid on the tray, and over the weir and into the downcomer to Tray 3. The vapor flows up from Tray 3 through holes in Tray 2, through the liquid flowing across Tray 2, disengages from the liquid, and continues to Tray 1. The vapor "bubbles" provide a large interfacial area for transfer of components between the phases. I put quotes around bubbles because the contact can be quite turbulent and even include a froth layer. A sieve tray (Figure 1.7b) simply has holes through which the vapor flows but they are often covered by bubble caps (Figure 1.7c) or valves (Figure 1.7d) that maintain better flow. Each tray is an equilibrium stage if 100% efficient, but that would be rare.

Figure 1.8a is a drawing of a column with random and structured packing. The liquid flows down the column, wetting the surface of the packing material. The vapor flows up the column through the void spaces inside and between the packing. Spreading the liquid over the packing surface provides a large interfacial area for transfer of components between the phases. Figure 1.8b shows a large variety of random packing designs and materials. Structured packing is shown in Figure 1.8c. Other internal structures are required to distribute and then redistribute the liquid as it flows down the column (Figure 1.8d) and support sections of packing. A certain height of packing is equivalent to a theoretical stage.

Figure 1.9a is a drawing of a condenser. It is a shell and tube heat exchanger in which the hot vapor enters the shell from the top, condenses on the surface of the tubes, and drains from the bottom. If possible, water is used in the tubes as coolant because of its low cost. Usually the condensate is collected in a reflux drum before being split into distillate and reflux. An industrial condenser is shown in Figure 1.9b.

A drawing of a kettle reboiler is shown in Figure 1.10a and of a thermosiphon reboiler in Figure 1.10b. Liquid is collected at the bottom of the column and flows to the reboiler. A tube bundle containing steam (usually) partially evaporates the liquid. The vapor is returned to the column as boilup and the liquid is taken as a bottom's product. If liquid is entrained in the vapor (intentionally in the case of the thermosiphon reboiler) it simply falls back into the pool of liquid at the bottom of the column. A picture of a kettle reboiler is shown in Figure 1.10c along with its tube bundle.

So now you have a qualitative picture of the processes we will learn to design. Most of this book will focus on "how tall should the column be?" That effectively means "how many equilibrium stages will be required?" Other questions include "what diameter column should be used" and "what are the best operating conditions?" We also need to know other information that chemical engineers might not be expert in, like the type of steel to use for the column, how thick it should be, how to properly weld it together, etc. No worries – separation processes are designed by teams of engineers and other specialists that collectively include all this expertise.

Psssst … I hope that you are curious about what chemical process equipment looks like and how it works. My first job out of college was to help build and operate a pilot scale nuclear fuel reprocessing pilot plant. It included a solvent extraction column in which uranium and plutonium were extracted from an aqueous solution of radioactive elements into an organic solvent. The column pressure was pulsed to aid in mixing the two liquids. That's a fascinating process and I wish that I could say I thoroughly studied solvent extraction in pulsed columns and the process chemistry. It would have served me well, especially whenever something went wrong. But in hindsight, I didn't learn nearly enough. Don't be like me! Become expert in everything you do. And while you're at it, why not keep a digital scrapbook of chemical process equipment schematics and pictures that you come across in your studies?

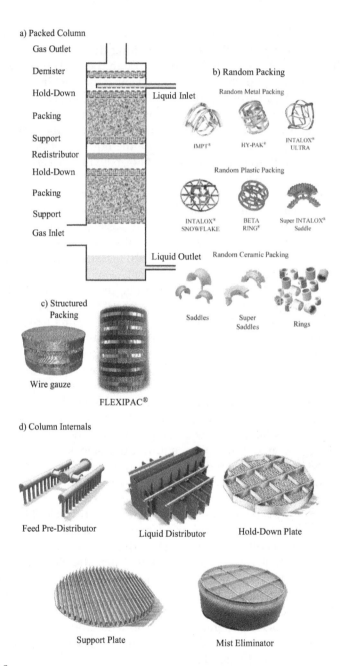

FIGURE 1.8
(a) Absorption column with random packing. (b) Various types and materials of random packing (1st row metal/2nd row plastic/3rd row ceramic). (c) Structured packing. (d) Various column internals. Images of metal and plastic packing, and column internals courtesy of Koch-Glitsch, LP, Wichita, KS, USA. Images of ceramic packing courtesy of MTE Group, The Netherlands. Images of structured packing courtesy of Sulzer Chemtech Ltd., Switzerland.

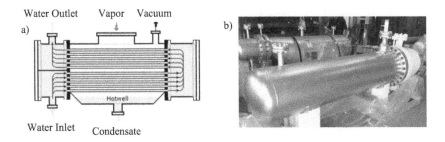

FIGURE 1.9
(a) Drawing of a total condenser, modified from Wikipedia, public domain. (b) Industrial condenser, courtesy of Koch Modular Process Systems, LLC, Paramus, NJ, USA.

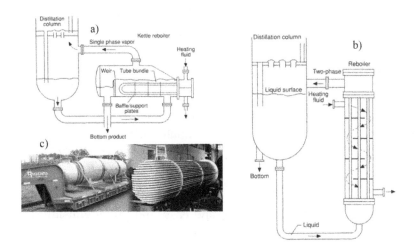

FIGURE 1.10
(a) Drawing of kettle reboiler. (b) Drawing of a thermosiphon reboiler. Both drawings reprinted from Hewitt, Geoffrey F., "Reboilers," Thermopedia, February 2, 2011. (c) Kettle reboiler used in industry with steam tube bundle exposed, courtesy of Koch Modular Process Systems, LLC, Paramus, NJ, USA.

Q: So, distillation and … what other processes will we look at?
A: We will focus on the "Big Three" separation processes: distillation, absorption/stripping, and solvent extraction. These all can be treated as equilibrium processes in which two phases are brought together, one or more chemical species transfer from one phase to the other, and then the phases are separated. We usually assume the separation has reached thermodynamic phase equilibrium and can correct that later. We'll stick to just a few examples for each type of process with additional examples presented in chapter problems at SeparationsBook.com. As an example of a nonequilibrium process, we'll also take a brief look at membrane separation.

Q: Can we assume there's a method to your madness?

A: Indeed – a highly structured method. For each process, we will examine just one or two chemical systems: benzene/toluene/xylene and ethanol/water for distillation; ethanol/water/air (or carbon dioxide) for absorption and stripping; water/acetic acid/isopropyl acetate (or MTBE) for solvent extraction; and oxygen/nitrogen and water/salt for membrane separations. Other systems will be introduced in chapter problems found at SeparationsBook.com.

Single equilibrium stages will be analyzed first by applying mole balances, phase equilibrium relationships, various process specifications, and, if necessary, energy balances. Most of the time, this results in multiple, coupled algebraic equations that must be solved together. Given enough process information, these equations will be solved in Excel spreadsheets to determine any unknown process variables. The only deviation from this will be analysis of membranes for which a transport equation replaces the phase equilibrium equation.

Then multiple stages will be added in series with countercurrent flow of the two phases, for example, liquid flowing down a column from stage to stage with vapor flowing up a column. The equations are the same except there will be a set for each stage. This is still solvable with Excel if the number and complexity of the equations are not too demanding. Always compare the solution with a process simulator like ChemCAD or Aspen, if possible.

You will notice a relationship between streams passing between stages and streams leaving a stage. This will lead you to "discover" convenient graphical methods for designing the separation processes. In many cases, these graphical methods are quick and easy, and provide insight into the process operation.

Q: What types of questions will we be asking?

A: That itself is a good question! Generally, we will either analyze or design a separation process. On one hand, we can specify a process: for instance, how many stages, which one will be the feed stage, and what are the process conditions (T and P)? Mathematical analysis – solving the equations – will then predict how the column will perform, what are the product flow rates and compositions. On the other hand, we can specify the performance, such as desired product flow rates and compositions. Graphical design will then determine how many stages are required and which one is the optimal feed stage. Of course, we can design with analytical methods or analyze with graphical methods, but the solutions require more trial and error.

1.1 Summary

Separation processes are ubiquitous in the chemical process industry. The most common ones are introduced in this book: distillation, absorption/ stripping, and solvent extraction. They can all be analyzed using equations you already know: material and energy balances, phase equilibrium relationships, and various process specifications.

2

A Look Inside Your Chemical Engineering Toolbox

Before we get started looking at separation processes, we need to review some basic chemical engineering tools. I assume that you have completed a material and energy balance course and a thermodynamics course. But the critical information in these courses that we will use to analyze and design separation processes came at you really fast and it has probably been a semester or two since you took them. The review here will not be exhaustive – just what you need to know for this course.

2.1 Process Flow Diagrams

Chemical engineers communicate using process flow diagrams (PFDs) as shown for the separator in Figure 2.1. At a minimum, they show the topology of the process: the major equipment and the pipes through which material flows from one unit to another. Generally, the process should flow left to right, top to bottom. PFDs don't have to be complicated. It is sufficient to use boxes to represent equipment and connecting lines to represent pipes. The geometric shapes can represent the function of equipment. For instance, use a vertical rectangle to represent a distillation tower, or perhaps a circle to represent a heat exchanger. Placement of lines should make sense; in this case, feed goes in the left side, vapor comes out the top and to the right, and liquid comes out the bottom and to the right. I've used color in the accompanying spreadsheet (Excel "Toolbox") but that is not necessary and sometimes not appropriate. In fact, the publisher told me "no color" in this book, so all the figures are black and white.

We can analyze the materials and energy that enter and leave a single-process unit, an entire complex process, or a subset involving several units. This is indicated by drawing a dotted line, an imaginary boundary, around the units indicating a material balance envelope (MBE). In Figure 2.1, there is just one process unit so drawing the envelope can be done but is actually unnecessary.

Many processes include several units, such as the distillation column pictured in Figure 2.2 that includes the column (C), total condenser (TC),

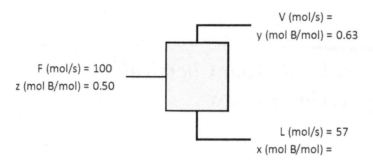

FIGURE 2.1
A PFD of a single-stage separator. This could be a partial evaporator (liquid feed), a partial condenser (vapor feed), or an adiabatic flash drum (liquid feed at a higher pressure). The labels include any specified information.

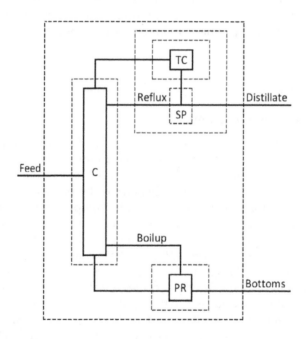

FIGURE 2.2
A PFD of a multi-unit distillation column. An MBE can be drawn around any unit or combination of units. The process units in this figure are column (C), total condenser (TC), split point (SP), and partial reboiler (PR).

partial reboiler (PR), a split point (SP) where a stream is divided, and a mix point (MP, not shown) where two streams come together. Many MBEs can be drawn. In fact, we will often draw MBEs around single or multiple stages inside the column. Only streams crossing the boundary are involved in a material and energy balance calculation.

2.2 Labeling

Information about the process is conveyed by labeling process properties near the equipment and streams. Label every stream so that all flow rates, compositions, and, if appropriate, intensive variables like temperature and pressure are known or represented by a variable. So long as there is no chemical reaction taking place, it doesn't matter if you express these in mass or mole units. I will usually use L, V, and F for the liquid, vapor, and feed molar flow rates; x, y, and z for their respective mole fractions; and T and P for temperature and pressure.

In the early chapters on vapor–liquid equilibrium (VLE) processes, I will use the benzene (B)/toluene (T) system in the examples. The labels in Figure 2.1 tell us that there are 100 mol/s fed to the process, including (100 mol/s)(0.50 mol B/mol) = 50 mol B/s of benzene. For a two-component mixture, the mole fraction of the other species, toluene in this case, is often left off since they simply add up to 1. Instead of total molar flow rate and mole fractions, we could also label the molar flow rate of each species: 50 mol B/s and 50 mol T/s.

Less common is to use molar ratios or concentrations. We could also say that we are feeding 50 mol B/h with a molar ratio of 1.0 mol T/mol B. Multiplying these together yields the molar flow rate of toluene. Another possibility is to use concentrations and volumetric flow rates, but this requires assuming constant density in material balance calculations and is best avoided.

If necessary, other properties should be identified. For instance, temperatures and pressures should be labeled if thermodynamic calculations are involved, such as vapor–liquid phase equilibrium and energy balances.

2.3 Material Balances

You probably perfected the use of this tool in the very first chemical engineering course. What goes in must come out or something else, like accumulation, happens. This is true whether we are talking about moles or mass of a substance. Oh sure, other things, particularly chemical reactions, can happen, but not in the separation processes that we will consider. In fact, most of the time we will assume steady-state operation so that material does not accumulate, so in = out. Sounds deceptively simple!

Well, you know that it can quickly get pretty complicated, especially with multi-unit processes. That is why engineers use a systematic process to solve material balance problems: (a) draw a labeled process flow diagram; (b) conduct a degree of freedom analysis (to see if the problem is solvable); and (c) write the equations and solve. This method is explained by

Felder, Rousseau, and Bullard (2015) and developed more fully in Chapter 3. I also add (d) explore the problem and (e) be awesome! to the process as you will see.

We are just counting moles in = moles out. The equations should be identified as they are written down.

$$\text{Total mole balance: } F = L + V \tag{2.1}$$

$$\text{Benzene mole balance: } Fz = Lx + Vy \tag{2.2}$$

The flow rates (F, L, V) have units mol/s and the mole fractions (x, y, z) have units mol B/mol. From Figure 2.1, we already know the values for F, z, L, y. These equations can be solved in the most efficient manner starting with any equations that have just one unknown. In this case, that is the total mole balance: $V = 100 - 57 = 43$ mol/s. With V now known, the benzene mole balance just has one unknown and can be easily solved: $x = \{(100)(0.50) - (43)(0.63)\}/(57) = 0.41$ mol B/mol.

Sometimes, two equations have to be solved simultaneously by elimination or substitution. But if the number of equations becomes large, they require iteration, or they need to be repeated often, they can be input to a spreadsheet. A surprising number of equations and unknowns can be handled by Excel's Solver tool. You will see plenty of examples in upcoming chapters so make sure you are reasonably skilled at using Excel spreadsheets.

Note that the vapor and liquid are usually assumed to be in equilibrium, meaning that other constraints on the products exist. As specified above, 57 mol/h of liquid and 0.63 mol B/mol in the vapor are probably not physically possible unless I made an awesome guess (or I happened to have worked it out ahead of time!). That's where phase equilibrium relationships come into the analysis.

2.4 Energy Balances

If the separator in Figure 2.1 is a partial evaporator, it will raise the temperature and vaporize some of the feed. Energy needs to be added, usually in the form of heat. Energy is counted in much the same way as moles/mass of the species: in = out. Energy enters with the feed and leaves with the products. It can also be transferred across surfaces because of temperature differences as heat, or imparted by moving parts (pumps, compressors, agitators, etc.) as shaft work. Because most separation processes involve flow of material, we will use enthalpy as the state function ($\hat{H} \equiv \hat{U} + P\hat{V}$), where $\hat{H}$ is the specific enthalpy (kJ/mol), $\hat{U}$ is the specific internal energy (kJ/mol), P is the pressure (bar), and $\hat{V}$ is the specific volume (m^3/mol).

The steady-state enthalpy balance equation is then

$$\Delta \dot{H} = \dot{H}_{out} - \dot{H}_{in} = \sum_{out} \dot{n}_i \hat{H}_i - \sum_{in} \dot{n}_i \hat{H}_i = \dot{Q} - \dot{W}_s \qquad (2.3)$$

where $\Delta \dot{H}$ is the difference in the product and feed enthalpies (kJ/s), $\dot{H}$ is the total flow of enthalpy of the feed or products (kJ/s), $\hat{H}$ is the specific enthalpy (kJ/mol), $\dot{n}$ is the molar flow rate of any species (mol/s), $\dot{Q}$ is the flow of heat (kJ/s) due to temperature differences, and $\dot{W}_s$ is the shaft work (kJ/s) due to moving parts. The overhead dot just means a rate and we will often drop it for convenience. Our processes will not include calculation of shaft work.

One of the most difficult and important energy concepts is that the absolute enthalpy of a material cannot be known. We can only know the change in enthalpy between two conditions or thermodynamic states (temperature, pressure, composition, and phase), for instance between one of the process stream conditions and a reference state. The reference state is arbitrary and chosen for convenience unless enthalpy is found from an equation or table that already is based on a specific reference state. A common example is the enthalpies given in steam tables that usually use the triple point of water as the reference state. Enthalpies from the National Institute of Standards and Technology (NIST) use liquid at the normal boiling point as a reference state.

Energy balance calculations will be more fully developed and demonstrated in Chapter 4. If you have not had a course in thermodynamics – or never really understood it – you can still master most of the material in this book.

2.5 Phase Equilibrium

We usually assume that two phases leaving a stage are in thermodynamic phase equilibrium. Each species will have a certain algebraic relationship between the compositions in each phase. For VLE, we write for each species, i,

$$y_i P \Phi_i = x_i p_i^* \gamma_i \qquad (2.4)$$

where y_i is the vapor-phase mole fraction, P is the pressure (and the product $y_i P$ is the partial pressure), Φ_i is vapor-phase fugacity coefficient, x_i is the liquid-phase mole fraction, p_i^* is the vapor pressure, and y_i is the liquid-phase activity coefficient. Φ_i and γ_i account for non-ideality in the vapor and liquid, respectively. Except at high pressures or low temperatures, the vapor is comparatively sparse ($\sim 10^3$ times lower density that the liquid) and behaves much like an ideal gas, with $\Phi_i \sim 1$. If a liquid also behaves ideally, the activity coefficient $\gamma_i \sim 1$ (this is basically the definition of an ideal liquid!) and we have Raoult's law

$$y_i P = x_i p_i^*$$
(2.5)

that is often written

$$y_i = K_i x_i \qquad K_i \equiv \frac{p_i^*}{P}$$
(2.6)

The benzene–toluene system behaves like an ideal liquid because they are similar chemicals.

The water–ethanol system is not ideal and even forms an azeotrope. We must describe the VLE with a liquid activity coefficient.

$$y_i = K_i x_i \qquad K_i \equiv \frac{p_i^* \gamma_i}{P}$$
(2.7)

There are a number of activity coefficient models to cover this situation and we will use the Wilson equation to calculate γ_i. The activity coefficient for a two-component system is a function of both liquid-phase mole fractions

$$\ln \gamma_i = -\ln\left(x_i + x_j \Lambda_{ij}\right) + x_j \left(\frac{\Lambda_{ij}}{x_i + x_j \Lambda_{ij}} - \frac{\Lambda_{ji}}{x_j + x_i \Lambda_{ji}}\right)$$
(2.8)

where Λ_{ij} and Λ_{ji} are Wilson coefficients for the two species. This equation does a good job of modeling ethanol–water VLE but complicates the calculations. Figure 2.3 shows the experimental VLE data as an xy graph and the fit for some common activity coefficient models. Raoult's law does a particularly poor job of describing ethanol–water VLE.

We often encounter gases dissolved at very low concentrations in liquids – think of carbonated or aerated water. It doesn't make sense to talk about their vapor pressure since they do not exist as a liquid under normal conditions. In this case, equilibrium is described by Henry's law with an experimentally determined and temperature-dependent constant, H_i.

$$y_i = K_i x_i \qquad K_i \equiv \frac{H_i}{P}$$
(2.9)

Solvent extraction involves separation of chemicals into two liquid phases. At equilibrium

$$\gamma_i^E x_i^E = \gamma_i^R x_i^R$$
(2.10)

where the superscripts E and R stand for the products "extract" and "raffinate," respectively. If the liquids separate (as they must for solvent extraction to work), the mixture is highly non-ideal and the activity coefficient must be determined in each liquid for each chemical. These systems are often modeled using the NRTL (non-random, two-liquid) equation. We will explore this type of phase equilibrium in Chapter 17 on solvent extraction.

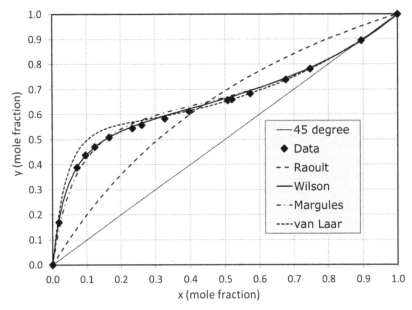

FIGURE 2.3
VLE data (♦) for ethanol–water mixtures showing the vapor-phase mole fraction (y) in equilibrium with the liquid-phase mole fraction (x) at 1 atm. This is called an xy graph. Various models are used to fit the data. The Wilson equation does best while Raoult's law is a very poor fit for this non-ideal system.

IS IT A VAPOR OR A GAS?

These terms are sometimes used interchangeably by scientists and engineers alike. The difference is more than just semantic. From Wikipedia, "In physics a vapor ... is a substance in the gas phase at a temperature lower than its critical temperature, which means that the vapor can be condensed to a liquid by increasing the pressure on it without reducing the temperature." Let's agree on a practical understanding. A vapor is in equilibrium with the liquid if it is saturated or could be if it is superheated and cooled to the dew point. So in distillation, we speak of VLE. A gas will not condense at any "normal" temperature although it can dissolve into a liquid. So in absorption/stripping, which usually involves a non-condensable component, we usually talk about a gas phase. Of course, it is not always straightforward, hence the confusion. What about absorption of ethanol from air using water? The ethanol can condense and the water can evaporate, so wouldn't they be vapors? What about cryogenic separation air into oxygen and nitrogen? Doesn't that involve VLE? Oh, my head hurts! Distillation – vapor. Absorption – gas. End of discussion.

2.6 Spreadsheets

More than any other software tool, chemical engineers use spreadsheets for numerical calculations and data presentation. Microsoft's Excel is by far the most common. Most of the problems we encounter in the analysis and design of separation processes involve solving multiple equations and often it is an iterative ("guess and check") process. It can take hours to solve a problem by hand. But if you set up the equations in Excel, they can be solved as fast as you can hit the "enter" key. This book will make extensive use of spreadsheets that are free downloads at SeparationsBook.com and referred to as Excel "Name/Page."

You are no doubt familiar with the use of spreadsheets, but I encourage you to become expert. After decades of using Excel, I am still learning new capabilities – often from my students! This will not be a comprehensive tutorial, just a few examples. Using the separator described in Figure 2.1, we can draw a labeled PFD (in fact, which is how I generated the figure) and write the equations with every term moved to one side so that they should equal zero if correctly solved.

Figure 2.4 shows the labeled PFD copied from Excel "Toolbox/Single Unit." The given data are inputted in their proper places (F, z_B, y_B, and L) and guesses (ridiculous ones at that!) are inputted as V (mol/s) = 100.0 and x_B (mol B/mol) = 1.000. The objective equations are the total mole balance (Total MB)

$$F - V - L = 0 \tag{2.11}$$

and the benzene mole balance (B MB)

$$Fz_B - Vy_B - Lx_B = 0 \tag{2.12}$$

The Solver tool is called up from the Data menu as shown in Figure 2.5. (You might have to add Solver from the File/Options/Add-Ins menu.)

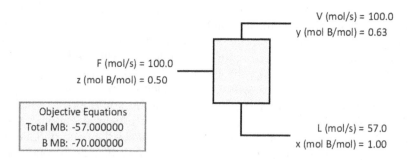

FIGURE 2.4
The PFD from Excel "Tool Box/Single Unit." The mole balances have been written so that all terms are on the left hand side (LHS) so they should be zero when satisfied. Guesses (and not good ones) have been made for V and x.

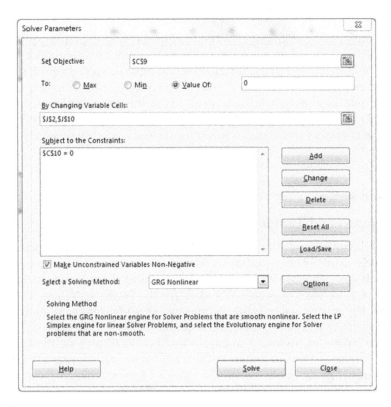

FIGURE 2.5
Solver menu from Excel "Toolbox/Single Unit."

The Objective (in this case the total mole balance in cell C9) and a Constraint (in this case the benzene mole balance in cell C10) are made equal to zero by changing the value for V in cell J2 and x_B in cell J10. I chose the unconstrained values to be nonnegative and the Simplex LP method. Pressing Solve yields the values in Figure 2.6.

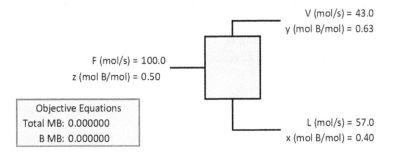

FIGURE 2.6
The PFD from Excel "Tool Box/Single Unit." Excel Solver has found values for V and x that satisfy the mole balances.

We just used Excel's Solver to solve two linear equations with two unknowns. You could have done this faster using pencil and paper. However, we will often encounter problems with many equations, some highly non-linear, and many unknowns that must be solved simultaneously by iteration. Solver uses highly efficient algorithms to search for the answers quickly and accurately. You have much better things to do than work these problems by hand, like going to the football game on Saturday. We'll have plenty of chances to use Excel's Solver tool in upcoming chapters.

2.7 Graphs and Tables

Chemical engineers (really, anybody!) can present information concisely and effectively using graphs and tables. The next time you pick up an engineering textbook, pay close attention to the format of these presentation tools. In this section, we'll generate VLE data for the benzene–toluene system at 760 mm Hg as another example of iterative spreadsheet calculations. Then we'll present that data in professional quality tables and graphs.

The calculations are found in Excel "Toolbox/VLE" and an abridged portion of the calculations is shown in Table 2.1. Each row contains a fixed value for the liquid-phase mole fraction, a guessed temperature, the vapor pressure of benzene and toluene calculated from Antoine's equation, the bubble point equation (derived from Raoult's law) set to zero

$$K_B x_B + K_T (1 - x_B) - 1 = 0 \tag{2.13}$$

and the vapor-phase mole fraction calculated from Raoult's law

TABLE 2.1

Spreadsheet for Generating VLE Data for Benzene–Toluene at 760 mm Hg

x (mol B/mol)	T (°C)	P_B^* (mm Hg)	P_T^* (mm Hg)	BP	y (mol B/mol)
0.00	110.6	1,784	760	0.0000	0.000
0.01	110.2	1,762	750	0.0000	0.023
0.02	109.7	1,742	740	0.0000	0.046
0.03	109.2	1,721	730	0.0000	0.068
...					
0.97	80.7	774	298	0.0000	0.988
0.98	80.5	769	296	0.0000	0.992
0.99	80.3	765	294	0.0000	0.996
1.00	80.1	760	292	0.0000	1.000

$$y_B = K_B x_B \qquad (2.14)$$

The correct temperature is found by iteration. Temperature is guessed; the vapor pressures are calculated and used to see if the bubble point equation is zero. This is repeated until the bubble point equation is satisfied. The vapor-phase mole fraction is calculated with each iteration, but the final value is correct. The resulting temperature and vapor-phase mole fraction correspond to the liquid-phase mole fraction in this row. That iteration can be done with the spreadsheet functions Data/What If Analysis/Goal Seek and repeated row by row. However, it can be done all at once using the Solver tool as seen in Excel "Toolbox/VLE." If your computer/software can't handle 100 rows of calculations, simply break the process into groups of rows.

The equilibrium data can be presented in either tables or graphs. Use the formats in your favorite engineering and technical writing textbooks to guide your presentation. Styles differ in details but general principles of effective communication are universal.

Table 2.2 is a good presentation of the data for a report. Only 10 points were included to make a more palatable reading of the table. Don't make long tables just because your spreadsheet generated them unless you are preparing a reference for people to extract data (for example, the steam tables). I have seen reports with tables containing four pages of numbers in very small font with eight "significant" figures simply because the author used small increments in their calculations and the computer spit them out. Please don't do that to your poor reader!

Note the general format. Tables always have a title at the top. In this case, it is "Table 2.2 Vapor–Liquid Equilibrium for Benzene–Toluene at 760 mm Hg." Each column has a heading that includes the units where appropriate.

TABLE 2.2

VLE Data for Benzene–Toluene at 760 mm Hg

x (mol B/mol)	y (mol B/mol)	T (°C)
0.000	0.000	110.6
0.100	0.209	106.1
0.200	0.376	102.1
0.300	0.511	98.5
0.400	0.622	95.1
0.500	0.714	92.1
0.600	0.791	89.3
0.700	0.856	86.8
0.800	0.911	84.4
0.900	0.959	82.2
1.000	1.000	80.1

Don't complicate the table with grid lines unless you have many rows or columns that really need a visual guide. You can create the table in Excel (as I did) and copy/paste it into your document or generate the table in Word.

An even better way to present this data is with a graph. Highlight the columns you wish to graph – perhaps an xy or a Txy graph – and use the "Insert/Scatter Chart" function to create the rough graph. Move the graph to a new page. It will look terrible and require extensive formatting. I actually formatted Figure 2.7 a little so it shows up better, but still has serious flaws.

In order to get a great looking xy graph like Figure 2.8 you should:

- Remove the title from the top – it is there mostly for PowerPoint presentations. Information will be in the caption. The caption will appear at the bottom in sentence form "Figure 2.9 Vapor–liquid equilibrium for benzene–toluene at 760 mm Hg. ...." Additional information can be included here or just in the text. Enough information should be included that a reader understands what they are looking at.

- Include axis labels with units to show what is plotted.

- Make all fonts large enough to read easily after it has been copied, pasted, and *resized* into your document.

- Use a reasonable number of scale markings. There is no need to include a physically impossible number like 1.2 mole fraction.

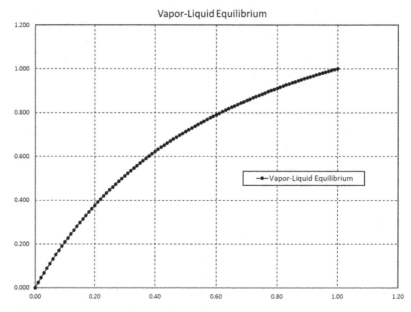

FIGURE 2.7
A very poorly designed xy graph. Specific problems discussed in text.

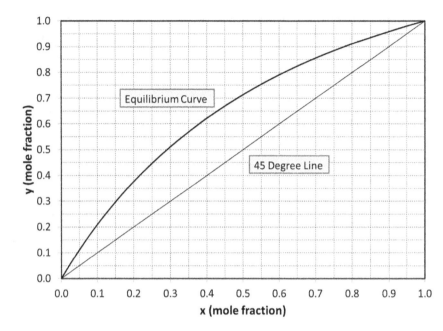

FIGURE 2.8
VLE for benzene–toluene at 760 mm Hg. This is called an xy graph.

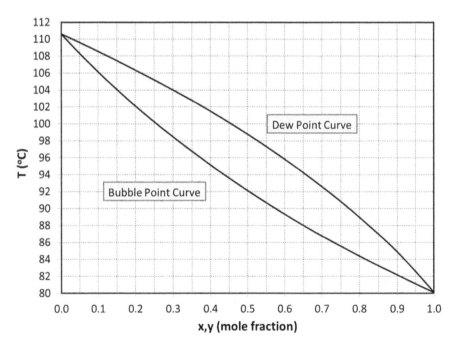

FIGURE 2.9
VLE for benzene–toluene at 760 mm Hg. This is called a Txy graph.

- Remove excess zeros on the axis values. These do not represent significant figures. So even if the mole fraction is 0.105, the scale should only read 0.1, not 0.100.
- Use markers for measured data with perhaps a fitting line. These numbers were all calculated so just use a line.
- Use grid lines if you expect your reader to extract values from the graph. If you are just showing trends, there is no need for the grid lines.
- Don't include a legend if there is just one line.
- Remove the box that surrounds the graph since it adds nothing to the presentation.
- You can use color if appropriate but consider if the graph will actually be printed in color. Otherwise distinguish lines with dashes and dots (Figure 2.3) or text box labels (Figures 2.8 and 2.9).

The formatted Txy diagram is shown in Figure 2.9.

The main point is that you should format these presentation tools to make it as easy as possible for your reader to understand your message. It takes a lot of time and thought but is an essential skill in a complete chemical engineering toolbox.

2.8 Computer-Aided Design Software

Every analysis and design problem we solve in this book is more easily solved using commercial chemical process computer-aided design (CAD) software such as ChemCAD or Aspen. In fact, as a professional chemical engineer, you will likely turn to these programs first to design separation processes. But it is essential that you understand what these programs are doing behind the scene in order to interpret the results. What are the material balance, energy balance, and VLE calculations involved? If you have access to CAD software, I encourage you to check the results of every example and chapter problem. You will experience some satisfaction knowing that you can come up with the same answers on your own.

2.9 Work Ethic

Your greatest superpower is your ability to work hard and smart. Learning engineering takes a lot of practice. There actually aren't that many things to memorize. You don't need to study for an exam using flash cards. But you do need to work problems – lots of problems – as many problems as you have the time and energy to conquer. If you coasted through high school and freshman year on your excellent memory and quick recall, you no doubt learned this was not sufficient for your introductory chemical engineering courses. Most engineering exams are time-constrained in order to challenge you and guarantee a grade distribution. You will do better if an exam is not the first time you encountered and solved a particular type of problem and if you have already solved as many variations of a problem as you can think of. This book is designed to help you do just that. So resist the temptation of midweek parties or staying up all night playing video games. Prepare as if this is your future career. Oh, that's right, it is!

Many students ask if this is what chemical engineers do for a living. Do you sit all day in a cubicle with a pile of homework problems and a computer? Absolutely not! No employer wants to pay you a professional salary to do that! However, when you are called on to quickly diagnose a malfunctioning distillation column costing your company a million dollars a day, or design a gas absorption column with a multidisciplinary team that considers you the expert, you will be glad that you developed a fundamentally sound understanding of separation processes.

2.10 Summary

Most chemical separation processes can be analyzed and designed using just a few types of algebraic equations: mole balances, energy balances, phase equilibrium relationships, and various process specifications. That's it! However, the equations are often non-linear and coupled, requiring iterative solutions. The calculations are often best performed with a spreadsheet. However, all your work is meaningless if you can't communicate your results. Fortunately, the spreadsheet is also an excellent tool to prepare PFDs, graphs, and tables. Your toolbox is complete. Are you ready to analyze and design some separation processes? Let's go!

Part II

Distillation

Most separation processes involve contacting or creating two phases, such as a liquid and vapor. Separation occurs as the chemicals distribute between the two phases up to the point of thermodynamic equilibrium. By far, the most common method is distillation. A liquid is fed and a vapor is produced by adding heat. "Light" compounds (lower boiling point) are concentrated in the vapor and "heavy" compounds (higher boiling point) are concentrated in the liquid.

To focus on fundamentals, benzene–toluene mixtures will be used in most examples with other applications introduced in the website's chapter problems (SeparationsBook.com). It is an industrially important separation and the phase equilibrium behavior is ideal so that Raoult's Law can be used.

I use 100 mol/s as the feed rate in most examples for convenience. For an equimolar mixture of benzene and toluene, that amounts to a little over 2,400 bbl benzene/day, which is in the range of a typical refinery capacity. Just for fun, try this unit conversion. Fun Fact: "barrel" has only one "b" so where does the extra "b" in the abbreviation come from? Standard Oil began producing blue-colored wood barrels around 1870 to designate a 42-gal barrel as opposed to the standard 40-gal barrel used in other industries. The extra two gallons were to provide for leakage and evaporation. These were known as blue barrels (US Energy Information Agency).

In Chapter 3 "Single-Stage Distillation: Material Balances," we will use just mole balances and vapor–liquid equilibrium (VLE) relationships to analyze partial evaporators and condensers. You will learn mathematical methods (solving the equations directly) and graphical methods (using VLE graphs).

We learn in Chapter 4 "Single-Stage Distillation: Energy Balances" how to determine heat requirements. Most of the time, we will calculate enthalpies (because these are flow systems) using latent (phase) and sensible (temperature) heat changes from a reference condition. We will also use published thermodynamic data from NIST. The amount of heat added to a partial evaporator or removed from a partial condenser can be determined. Also, the question can be flipped – given a specified amount of heat, how does a separator perform? In this way, an adiabatic flash drum can be analyzed for which the amount of heat transferred is zero by definition.

In Chapter 5 "Multi-Stage Distillation," we learn that the very same material balance and VLE equations we used to analyze a single equilibrium stage can be used for multiple stages stacked in series. Excel spreadsheets are introduced early in Part II to solve these simple equations. However, the algebra starts to get a little complicated because of the number of equations in these multi-unit systems, so that Excel spreadsheets become an essential tool. We will find that single equilibrium stages are limited in how much separation can be achieved. But multiple equilibrium stages in series produce much better separation.

In Chapter 6 "Mathematical Analysis of Distillation Columns," we will again apply the very same material balance and VLE equations to entire distillation columns with ten stages and solve them with the efficient Lewis method. This is an analysis problem – given the design of a distillation column (number of stages, location of the feed stage, operating conditions, etc.) what is the performance (product flow rates and compositions, etc.)? In Chapter 7 "Graphical Design of Distillation Columns," we develop the McCabe–Thiele technique that uses a simple construction on the xy graph to solve design problems – given the desired performance of a distillation column, how should the column be designed?

The techniques learned in Chapters 6 and 7 do not require knowledge of the energy balances around any or all stages. In Chapter 8 "Energy Balances for Distillation Columns," we return to the very same energy balance equations introduced in Chapter 4 and add them to the analysis. How much heat must be added to the partial reboiler? How much heat must be removed from the total condenser? Also, the Lewis (Chapter 6) and McCabe–Thiele (Chapter 7) methods required the assumption of equimolal overflow (EMO) – that the liquid and vapor molar flow rates are constant above and below the feed stage. Most students find this assumption a bit mysterious, but it can be dropped if we include a stage-by-stage energy balance.

Chapter 9 "Distillation: Variations on a Theme" addresses some (of course, not all) of the many variations and complications that are found in distillation processes. What if the actual design is different, including taking multiple product streams, feeding to more than one stage, or adding steam directly to the bottom of the column? You will find that the same systematic approach to problem-solving that you used in earlier chapters can be applied to any of these questions.

Chapter 10 "Multicomponent Distillation" introduces three- and four-component distillations. Benzene and toluene streams usually contain other components such as the three xylene isomers and ethylbenzene. Finally,

Chapter 11 "Distillation of Non-Ideal Systems" shows how to analyze systems that do not obey Raoult's Law. The example chemicals in this case are ethanol and water that form an azeotrope that can be predicted with Wilson's equation for the activity coefficient.

EQUIMOLAL OVERFLOW OR CONSTANT MOLAL OVERFLOW?

I've heard this assumption expressed both ways with constant molal overflow (CMO) seemingly the most common. I have always used equimolal overflow (EMO) and it is hard to teach an old dog new tricks. They mean the same thing and EMO is used in the famous chemical engineering song "Steppin' Off The Stages." So there.

3

Single-Stage Distillation: Material Balances

We can partially separate a mixture of chemicals with different volatilities by feeding a single phase (liquid or vapor) to a vessel in which temperature and/or pressure is changed to create a second phase. The chemicals distribute between the two phases, approaching thermodynamic equilibrium, so this is called a single equilibrium stage. The vapor is removed from the top and is more concentrated in the "light," more volatile component. The liquid is removed from the bottom and is more concentrated in the "heavy," less volatile component. A mixture of benzene (light) and toluene (heavy) will be used as an example since it is an "ideal" mixture with vapor–liquid equilibrium (VLE) that can be described by Raoult's law. Some example processes are:

1. Partial evaporator: a liquid mixture is fed and heat is added to partially evaporate the liquid to create a vapor.

2. Partial condenser: a vapor mixture is fed and heat is removed to partially condense the vapor to create a liquid.

3. Adiabatic flash: a liquid mixture at a relatively high pressure is fed to a "flash drum" that operates at a lower pressure. It is well insulated so that no heat is added or removed (an adiabatic process). The drop in pressure causes some of the liquid to vaporize provided conditions cross into the two-phase region. This analysis requires an energy balance and discussion is deferred to Chapter 4.

In this chapter and in all of Part II, you will learn how to use mathematical (i.e., using equations) and graphical methods to analyze or design VLE separation processes. We will stick to systems that create a vapor from a liquid (or vice versa) but very similar methods are used for other processes that add (instead of create) the second phase (absorption and stripping in Part III) or use other phase equilibria (solvent extraction in Part IV).

Most topics will be introduced with a short, general technical discussion. This will be followed with a worked example and then a guided "Try This at Home" example in which you do most of the work. More examples with a variety of chemical mixtures are in the chapter problems at SeparationsBook.com. You really should work as many problems as possible. The late UA Professor David Hart sat at his kitchen table every night (even after he retired!) and worked textbook problems. He said, "In order to learn engineering, you must do engineering." So true!

3.1 Systematic Solution Procedure

In all example problems, we will follow a systematic solution procedure similar to that described by Felder, Rousseau, and Bullard (2016):

a. Draw a labeled process flow diagram.

First draw and label a process flow diagram (PFD) with all known flow rates and compositions (e.g., 100 mol/s, 0.50 mol B/mol) and use algebraic symbols for unknowns (e.g., V (mol/s), y_B (mol B/ mol)). The time unit "dot-above" will be often left off for convenience so that V is understood to mean $\dot{V}$ (mol/s). Also, one mole fraction is often not labeled since the mole fractions are understood to add up to 1.

b. Conduct a degree of freedom analysis.

Next perform a degree of freedom (DOF) analysis to make sure the problem is properly specified so that the number of equations matches the number of unknowns.

c. Write the equations and solve.

Only then should you write down the equations and solve. These equations will include material balances, equilibrium relationships, and various process specifications. Often it will be most convenient to use a spreadsheet to solve multiple equations for multiple unknowns by iteration. Graphical techniques can sometimes be used instead of algebraic equations.

If the heat added or removed from a process is required, an energy balance equation will be included. If the heat is specified, the energy balance and material balance equations will need to be solved simultaneously.

d. Explore the problem.

Once you have gone to the trouble of solving one of these complicated problems, don't just stop there. Explore the parameters to see what happens to the problem solution and explain your results. This is particularly easy to do if you use a spreadsheet solution. Instructions and recommended experiments are included in every spreadsheet.

e. Be awesome!

OK – I added this one! You are awesome for having the audacity to even study chemical engineering (see my YouTube talk "The Audacity of Awesomeness"). Take these solutions to the next level by, for instance, modifying a spreadsheet to solve a different type of problem or creating your own spreadsheet solution using the accompanying DIY template. Suggestions will be included in the examples.

3.2 Mathematical Methods

Suppose two chemicals are fed to a process as a single phase and two product streams, a vapor and a liquid, are produced. We are given certain information and asked to calculate all remaining flow rates, compositions, and intensive variables like temperature and pressure. This process involves a phase change that requires large exchanges of energy; however, we will postpone the energy calculation until Chapter 4. So let's begin!

3.2.1 Mole Balances

The process variables (flow, composition, phase, temperature, and pressure) are related by just a few fundamental equations *that you already know*. The total mole balance equation simply states that the total number of moles coming into the process equals those leaving (assuming steady state and no reaction).

$$F = L + V \qquad (3.1)$$

F is the feed, L is the liquid product, and V is the vapor product (all in units mol/s). The species mole balance equation is similar but for an individual chemical.

$$Fz_i = Lx_i + Vy_i \qquad (3.2)$$

x_i, y_i, and z_i are mole fractions (mol i/mol) of species "i" in the liquid, vapor, and feed streams, respectively. We can write one mole balance for each species present. So for a two-component system, we can write the total mole balance and one species mole balance or two species mole balances. The third mole balance equation would not be an independent equation.

3.2.2 Phase Equilibrium

We will also specify that the two product phases are in VLE. The temperature (thermal equilibrium) and pressure (mechanical equilibrium) in each product stream will be the same ($P_L = P_V$, $T_L = T_V$). Also the compositions can be related, commonly by Raoult's law for ideal mixtures.

$$y_i P = p_i^* x_i \qquad (3.3)$$

P is the total pressure and p_i^* is the vapor pressure. We will often write Raoult's law as

$$y_i = K_i x_i, \qquad K_i \equiv \frac{p_i^*}{P} \qquad (3.4)$$

The vapor pressure is most commonly calculated using Antoine's Equation, which comes in many forms (so pay attention to how it is defined). The appendix uses the form

$$\log_{10} p^* = A - \frac{B}{T+C} \tag{3.5}$$

where A, B, and C are constants, p^* is the vapor pressure (mm Hg), and T is the temperature (°C). In some resources, it might be written with a natural log and different P and T units. For instance, the National Institute of Standards and Technology recommends bar and K.

So, that's it?! Well, yes! Other than miscellaneous process specifications, these are all the equations we need. If the temperature is not specified this becomes a trial and error solution and best left for a computer solution. We'll use Excel's Solver tool for many example calculations. So let's give this a try!

EXAMPLE 3.1   Partial Evaporator – Mathematical Method

A total of 100 mol/s of a liquid mixture with 40 mol% benzene and 60 mol% toluene is fed to a partial evaporator operating at 760 mm Hg. The vessel is heated to vaporize a fraction of the liquid and operates at VLE. Determine the evaporator's temperature and the amount and composition of each phase.

a. Draw a labeled process flow diagram.

   A PFD is shown in Figure 3.1. Note that the mole fraction of toluene is left off because the mole fractions simply add up to 1. We know the PFD is properly labeled because all flow rates and mole fractions are either known or represented by a variable. If we specify the products are in VLE, it is necessary to also include temperature and pressure in the labeling for use in Raoult's law.

b. Conduct a degree of freedom analysis.

   Before jumping into the calculations, we should ask if there is enough information given to solve the problem. This is the purpose of a DOF analysis. Count the number of unknowns and the number of equations or relationships between these variables. If they are equal, we can solve the algebraic equations. If there are more equations, the problem is over-specified. If there are more unknowns, the problem is underspecified.

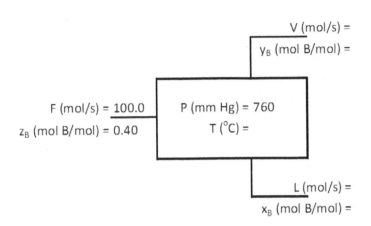

FIGURE 3.1
A partial evaporator PFD with process variables identified and some values noted from the process description in Example 3.1.

Unknowns (5): V, y_B, L, x_B, and T

Equations (4):

2 mole balances (total and B)

2 equilibrium relationships (Raoult's law for B and T)

So that's 5 unknowns and 4 equations – one DOF – which means the problem is underspecified. That can actually be good news for you, the engineer, because you have some flexibility in the design.

We need to specify one more piece of information. Some possibilities are:

1. Specify one more variable: V, y_B, L, x_B, or T. Note that V and L must be less than 100 if we are to actually have two phases. Also, choices of T, x, and y are limited for a system at equilibrium. This is easiest to see using xy and Txy graphs.

2. Specify the fraction vaporized (V/F), the fraction remaining as a liquid (L/F), or the ratio of the two phases (V/L). Note that V/F and L/F must be between 0 and 1, while V/L must be between 0 (all liquid, no vapor) and ∞ (all vapor, no liquid).

3. Specify a percentage recovered, such as 60 mol% of the benzene is recovered in the vapor. Choices are also limited for a system at equilibrium.

For this example, we will specify that the vapor flow rate is twice the liquid flow rate. Now we have zero DOF and can proceed to write the appropriate equations.

c. Write the equations and solve.

Be sure to identify each equation.

Total mole balance: $F = V + L$

Benzene mole balance: $Fz_B = Vy_B + Lx_B$

Benzene equilibrium: $y_B = K_B x_B$ (remember that $K_B = p_B^*/P$ for an ideal solution and p_B^* is determined from Antoine's equation)

Toluene equilibrium: $(1 - y_B) = K_T(1 - x_B)$ (written in terms of benzene since we didn't label toluene)

Process specification: $V = 2L$

where F, V, L [=] mol/s and x_B, y_B, z_B [=] mol B/mol.

We already know the values for F, z_B, and P. If T is unknown (as in this example) a trial and error solution of these five equations is time-consuming because of the temperature in the vapor pressure (Antoine) equation. Solving these equations using a spreadsheet is quick and easy using Excel Solver. Excel "Single Stage/1-Stage" determined the following values: $V = 66.7$ mol/s, $L = 33.3$ mol/s, $x_B = 0.265$ mol B/mol, $y_B = 0.467$ mol B/mol and $T = 99.7°C$.

Go ahead and pencil these values into Figure 3.1. It's OK – I encourage you to write in your book!

The evaporator increased the temperature and created a vapor phase requiring large amounts of sensible and latent heat to be added. This will be calculated using an energy balance in Chapter 4.

d. Explore the problem.

Check out Excel "Single Stage/1-Stage" and make sure you understand the equations in every cell. Put guessed values in the cells for V, L, x_B, y_B, and T and run Solver. You should find the above results. Use these values to demonstrate with hand calculations that they satisfy the above equations. Try other specifications. The Excel spreadsheet contains detailed instructions and suggested experiments.

Notice that we didn't specify the complete thermodynamic state (T, P, and phase) of the feed. The outlet flow rates and compositions only depend on the evaporator's T and P and the flow rate and composition of the feed as long as we are not asking how much heat needs to be added or removed.

e. Be awesome!

Surely you are curious about what these processes look like in industry. Do a little investigation on the Internet to see what you can find. Start a digital collection of chemical process equipment schematics and pictures.

A QUICK NOTE ABOUT HAND CALCULATIONS

You will notice that I frequently challenge you to determine or verify values by hand calculations, meaning working with pencil, paper, and a calculator. It is one thing to read and understand an equation, but quite another to actually put pencil to paper and do the calculation yourself. It is an active, as opposed to passive, form of learning. You should become expert in these calculations and learn to solve equations quickly and accurately. If you are well practiced on these routine and mundane calculations, you will have that much more time on an exam to solve more complex and nuanced problems.

**TRY THIS AT HOME 3.1   Partial
Condenser – Mathematical Method**

A total of 100 mol/s of a 106°C vapor mixture with 40 mol% benzene and 60 mol% toluene is sent to a partial condenser operating at 760 mm Hg and 99.0°C. Determine all unknown flow rates and compositions.

a. Draw a labeled process flow diagram.

The PFD is already partially done in Figure 3.2. Fill in what you know and leave the rest blank.

b. Conduct a degree of freedom analysis.

List the unknown flow rates and compositions and then identify the available equations. This is a properly specified problem, so they should be equal. You should identify four unknowns and four equations. At this point, just identify the equations – do not write them out. I'll start.

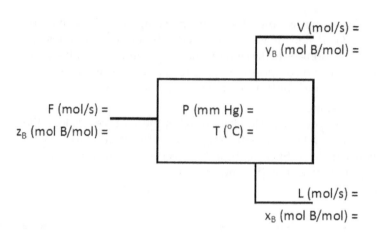

FIGURE 3.2
A partial condenser PFD with process variables identified for use in Try This at Home 3.1.

Unknowns:	Equations:
1. V (mol/s)	1. 1 total mole balance
2.	2.
3.	3.
4.	4.

If it looks like zero DOF, go on to the next step. If not, you are doing something wrong! If you are stuck, view my YouTube tutorial "Single Stage."

c. Write the equations and solve.
   Be sure to identify each equation. Again, I'll start.

1. Total mole balance: $F = V + L$

2.

3.

4.

Since T is given, you can easily calculate partial pressures from Antoine's equation (constants in the Appendix) and solve everything by hand, using a calculator, math software, or a spreadsheet. After inputting these new values, Excel "Single Stage/1-Stage" gives V = 55.6 mol/s, L = 44.4

mol/s, $y_B = 0.492$ mol B/mol, and $x_B = 0.285$ mol B/mol. Pencil these values into Figure 3.2. Verify with hand calculations.

The condenser lowered the temperature and created a liquid phase requiring large amounts of energy removed. This will be calculated using an energy balance in Chapter 4.

d. Explore the problem.

Use Excel "Single Stage/1-Stage" to check the answers for this particular problem and then try lots of other sets of values. Try V = 99.99 mol/s, almost all vapor, which is approximately the feed dew point. Likewise, try L = 99.99 mol/s, almost all liquid, which is approximately the feed bubble point. Use the Txy graph (Figure 3.3) and xy graph (Figure 3.4) to determine the ranges of possible T, x_B, and y_B. Use any value in this range as your specification. It's just a computer simulation – nothing will actually blow up! See what happens if you use a value outside this range and explain the result. Compare results with your study group. Take it to the next party and amaze your nonengineering friends! On second thought … maybe best to leave it at home!

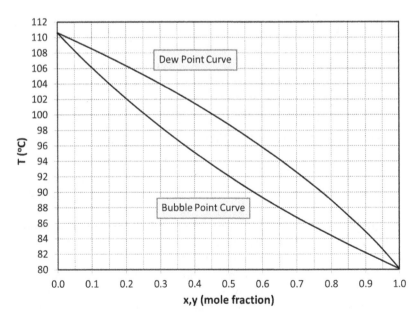

FIGURE 3.3
VLE for benzene/toluene mixtures at 1 atm. This is referred to as a Txy graph.

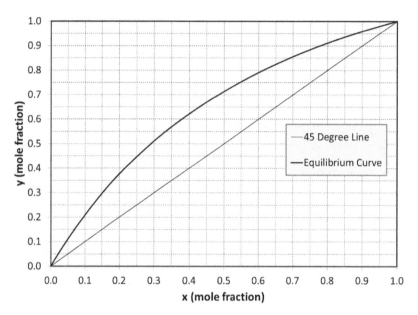

FIGURE 3.4
VLE for benzene/toluene mixtures at 1 atm. This is referred to as an xy graph.

　　e. Be awesome!
　　　　Make your own spreadsheet simulation. Practicing engineers need to be skilled at working with spreadsheets. A DIY template is provided with every spreadsheet.

DON'T TRY THIS AT HOME 3.1　Distilling Moonshine

Fermentation can only yield a limited alcohol content (say ~ 15% ABV (alcohol by volume) – but this depends on many factors) because it is toxic to the very yeast that produces it as a metabolic by-product. This is how beer, wine, and cider are made. In the 12th century, Italians and Chinese learned to distill fermentation products to make more concentrated "hard liquors" such as whiskey and vodka that have alcohol contents of ~40% ABV. Experts like the late Mr. Marvin 'Popcorn' Sutton (Figure 3.5) have perfected the art. Do a little Internet investigation and consider how this process works compared to the partial evaporator described above. Make a list of differences and similarities. See my tutorial YouTube "Distilling Moonshine" for my analysis.

FIGURE 3.5
Mr. Marvin 'Popcorn' Sutton (right) with a large moonshine still.

Differences	Similarities

Where weather permitted, high alcohol content drinks were made much earlier than the 12th century, by leaving barrels of the fermentation product outside to freeze and fishing out chunks of ice. This left most of the alcohol behind and, I suppose, made a refreshingly chilled drink! This process is called fractional freezing.

Not only is making homemade moonshine without a license illegal, the process can be hazardous and you are probably too young anyway! So don't try this at home!

3.3 Graphical Methods

Phase equilibrium for two-component systems like benzene and toluene mixtures is conveniently represented by the two-dimensional Txy (Figure 3.3) and xy (Figure 3.4) graphs. These can be used to obtain easy solutions to the material balances. Note that by convention, the mole fractions are for the most volatile component (benzene in this case). Generation of these graphs in Excel "Toolbox/VLE" was demonstrated in Chapter 2 but also repeated in most other spreadsheets, in this case Excel "Single Stage/VLE."

The Txy graph shows the temperature as a function of the liquid and vapor mole fractions at a specific pressure, in this case 1 atm. The curves are derived from the bubble point equation (bottom curve representing the liquid) and the dew point equation (top curve representing the vapor). The product vapor and liquid have the same temperature so, if temperature is known, simply draw a horizontal line at the given temperature on the Txy graph. It will intersect the bubble point curve at the liquid composition and the dew point curve at the vapor composition. If instead a composition is specified, drawing a horizontal line at that composition will yield the temperature and the other phase's composition. The Txy graph can even be used to calculate V and L using the "lever rule," demonstrated below and in my YouTube tutorial "Lever Rule."

The xy graph shows the family of all possible equilibrium vapor and liquid compositions at a specific pressure. The further the equilibrium line is from the 45° line, the larger the relative volatility and easier the separation. Each point on the curve is associated with a unique temperature but that information is lost in this type of plot.

TRY THIS AT HOME 3.2   Fun with Txy and xy Graphs

Pick any temperature between the benzene and toluene boiling points (T at x = 1 and x = 0, respectively) and draw a horizontal line through the two curves on the Txy graph, Figure 3.3. Note the liquid and vapor compositions. Find the liquid composition on the xy graph, Figure 3.4, and the vapor composition will be the same value that you found on the Txy graph. Try 88°C and you should find x ~ 0.650 and y ~ 0.825.

A line representing all possible solutions to the mole balance equations can be added to the xy graph and its unique intersection with the equilibrium line satisfies both material balance and equilibrium equations. The benzene mole balance is

$$Fz = Vy + Lx \qquad (3.6)$$

F, V, and L are the feed, vapor, and liquid molar flow rates and z, y, and x are their corresponding benzene mole fractions. Rearranging the equation into a straight line form, this becomes

$$y = -\left(\frac{L}{V}\right)x + \left(\frac{F}{V}\right)z \qquad (3.7)$$

This is called the operating line and will have much significance in the analysis of distillation columns. If x = z this equation becomes

$$y = -\left(\frac{L}{V}\right)z + \left(\frac{F}{V}\right)z = \frac{(F-L)}{V}z = \frac{V}{V}z = z \qquad (3.8)$$

The point $x = y = z$, or (z, z), is always on the operating line and also on the 45° line. If you know $-L/V$, which is the slope of the line, you have a point (z, z) and a slope allowing construction of the entire line. If you know one product composition, x or y, then with (z, z) you have two points that form a straight line.

Let's give this a try!

EXAMPLE 3.2   Partial Condenser – Graphical Solution

In this example, we'll use the same specifications as Try This At Home 3.1. 100 mol/s of a 40 mol% benzene/60 mol% toluene vapor is fed to a partial condenser operating at 1 atm and 99.0°C. Use the Txy and xy graphs to determine all unknown molar flow rates and compositions.

a. Draw a labeled process flow diagram and

b. Conduct a degree of freedom analysis.
   Refer to the previous PFD (Figure 3.2) and DOF analysis instead of repeating these here. This problem is solvable and the xy and Txy graphs represent the equations.

c. Write the equations and solve.
   The graphical solutions take the place of the mole balance and equilibrium equations. Print extra copies of the Txy and xy graphs from Excel "Single Stage/Txy and /xy" and follow along with the solution to this problem. Use the Txy graph in Figure 3.6 first because we know the temperature. Draw a horizontal line at 99.0°C because the liquid and vapor will have the same temperature (thermal equilibrium). The intersection with the bubble point curve is the liquid-phase composition, $x_B \sim 0.29$ mol B/mol. (Unlike the mathematical solution, it is hard to read more than two significant digits from this graph.) The intersection with the dew point curve is the vapor-phase composition, $y_B \sim 0.49$ mol B/mol.
   Use the xy graph in Figure 3.7 next because now we know the equilibrium compositions. Locate the feed composition on the 45° line (0.4, 0.4) and either the vapor or liquid composition on the equilibrium line. Draw a straight line through these points; that is, the operating or material balance line for this situation. The slope is $-L/V$ and can be used to determine the product molar flow rates. However, let's use the y-intercept instead (review Eq. 3.7).

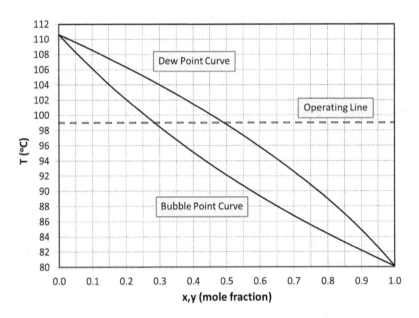

FIGURE 3.6
VLE for benzene–toluene at 1 atm showing the equilibrium vapor and liquid compositions at 99°C as described in Example 3.2.

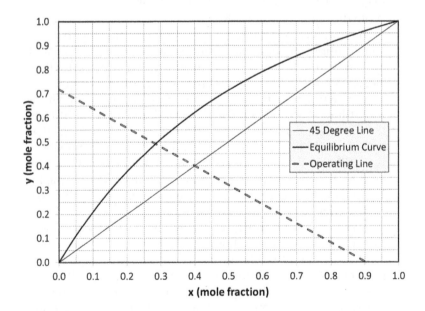

FIGURE 3.7
VLE for benzene–toluene at 1 atm showing the operating line as described in Example 3.2.

$$\text{Intercept} \sim 0.72 = \frac{F}{V} z_B = \frac{100}{V} 0.4, \ V = \frac{100}{0.72} 0.4 = 56 \frac{mol}{s}$$

$$L = F - V = 100 - 56 = 44 \frac{mol}{s}$$

Alternatively, we can use the inverse lever rule on the Txy graph as seen in Figure 3.8. The fraction that is liquid is represented by the length of segment $\overline{zy}$ divided by the length of segment $\overline{xy}$, L/F = (y − z)/(y − x) = (0.49 − 0.40)/(0.49 − 0.29) = 0.45, or L = (0.45)(100) = 45 mol/s, which is within round-off error of the mole balance above. The segment opposite of the bubble point curve represents the fraction of liquid phase – do you see why it is called the inverse lever rule? For vapor, the opposite is found: V/F = (z − x)/(y − x). Take a few minutes to derive these equations from the benzene mole balance.

If you input these specifications to Excel "Single Stage/1-Stage" and use Solver to find the mathematical solution, you will notice that the Txy and xy graphs are automatically updated with these operating lines. I hope you're thinking, now that's pretty cool!

The condenser decreased the temperature and created a liquid phase requiring large amounts of energy removed. This will be calculated using an energy balance in Chapter 4. How would you remove heat from the condenser? What would

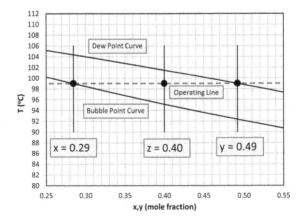

FIGURE 3.8
A region of the xy graph showing the liquid, feed, and vapor compositions at equilibrium.

some efficient designs look like? Brainstorm as many methods as possible.

d. Explore the problem.

Print several blank xy and Txy graphs from Excel "Single Stage/Txy and /xy." Specify F, z_B, P, and one more process variable: T, x_B, y_B, L, or V. Warning: not every temperature will put the mixture in the two-phase region. High temperatures will cause your computer to blow up! Just kidding. Use 760 mm Hg since the graphs are specifically calculated for atmospheric pressure. Use the graphs to determine the remaining process variables. Use Excel "Single Stage/1-Stage" to calculate these process variables and compare your answers. The operating lines are also automatically drawn on the spreadsheet graphs, allowing comparison with your graphical calculations. Rinse and repeat as often as necessary.

e. Be awesome!

If this material seems easy so far, it might be because you already saw it in the first chemical engineering course "Material and Energy Balances" (or some similar name). You have really nailed a topic when you can teach it. Find a willing nontechnical friend (or maybe one of your classmates) and see if you can teach single equilibrium stage, analytical and graphical methods to them.

**TRY THIS AT HOME 3.3   Partial
Evaporator – Graphical Solution**

In this problem, you will use the specifications in Example 3.1. A total of 100 mol/s of a liquid mixture with 40 mol% benzene and 60 mol% toluene is fed to a partial evaporator operating at 760 mm Hg. The product vapor flow rate is twice that of the product liquid flow rate. Use the Txy and xy graphs to determine all unknown molar flow rates, compositions, and the temperature.

a. Draw a labeled process flow diagram and

b. Conduct a degree of freedom analysis.

These steps were done previously in Example 3.1. You will not need to repeat the PFD and DOF analysis here. The problem is solvable.

c. Write the equations and solve.

Again, the graphical solution takes the place of the algebraic equations. Neither the temperature nor compositions are known so the Txy graph will not be of immediate use. Instead you can use the xy graph with one point (the feed) and the slope (–L/V) to draw the operating line. On Figure 3.9, locate the feed composition $z_B = 0.40$ on the 45° line. Next draw a line with slope $= -(L/V) = -0.5$ from that point. This operating line, representing the mole balance equations, intersects the equilibrium line at (x_B, y_B).

Locate these compositions on the Txy graph in Figure 3.10 and draw a horizontal line. Carry this line all the way to the vertical axis and read the temperature. Now you know the temperature, flow rates, and compositions of all streams.

Your analytical solution was $V = 2L = 66.7\,\text{mol/s}$, $T = 99.7°C$, $x_B = 0.265\,\text{mol B/mol}$, and $y_B = 0.467\,\text{mol B/mol}$. How does your graphical solution compare?

The evaporator increased the temperature and created a vapor phase requiring large amounts of energy added. This will be calculated using an energy balance in Chapter 4.

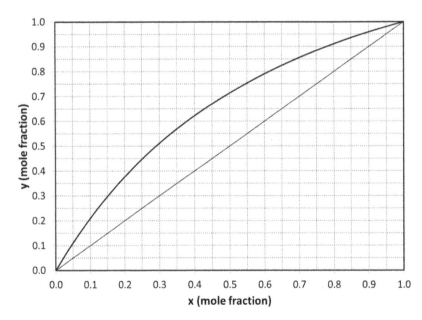

FIGURE 3.9
VLE for benzene–toluene at 1 atm for use in Try This at Home 3.3.

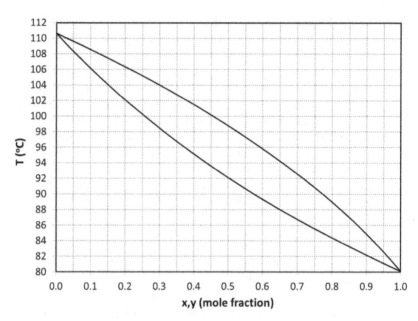

FIGURE 3.10
VLE for benzene–toluene at 1 atm for use in Try This at Home 3.3.

How would you add heat to the evaporator? What would some efficient designs look like? Brainstorm as many methods as possible.

d. Explore the problem.

Repeat part (d) from Example 3.2. Does it make a bit of difference if the single equilibrium stage is a condenser or evaporator? No, it doesn't as far as this analysis goes. But working more problems is never a bad idea.

e. Be awesome!

Make the xy and Txy graphs at a different pressure. Instructions for doing this are in Excel "Single Stage/VLE." How does using a different pressure change your solution? Discuss with your study partners.

3.4 Summary

A single equilibrium stage was described so that given sufficient information, any remaining unknown process variables could be calculated (flow rates, compositions, pressure, and temperature). A systematic procedure was demonstrated in which you drew a labeled process flow diagram, conducted a degree of freedom analysis, and wrote and solved the appropriate equations, including mole balances, VLE phase relationships, and other process specifications. As always, you took the initiative to explore the problems to squeeze out as much understanding as possible. That's awesome!

4

Single-Stage Distillation: Energy Balances

It seems obvious that a partial condenser will require heat to be removed, a partial evaporator will require heat to be added, and an adiabatic flash drum will by definition have no heat transfer. As long as we weren't asking questions about the required energy changes, the outlet conditions did not depend on the thermodynamic state (T, P, phase) of the feed; only the flow rate and composition mattered.

4.1 Calculation of Heat

So how much heat must be added to a partial evaporator or removed from a partial condenser? We use an enthalpy balance (since this is a flow system) that simply says the heat transferred, Q, is equal to the difference between the product and feed enthalpies, H,

$$Q = H_{out} - H_{in} = \sum_{out} n_i \hat{H}_i - \sum_{in} n_i \hat{H}_i \qquad (4.1)$$

where Q, H_{out}, and H_{in} [=] kJ/s, n_i [=] mol/s, and $\hat{H}_i$ [=] kJ/mol for each component. You might remember that there is no absolute value for specific enthalpy, only a value relative to a reference state *that we choose*. As long as we choose for each species a consistent reference state, the answer will always be the same. If you look up enthalpies in a table, be sure to note the reference state for that data and be consistent (see Example 4.3).

We will calculate enthalpy at a condition of interest (usually a feed or product stream) relative to a reference state by combining sensible and latent heat changes along a hypothetical path. If the reference state is a liquid, the enthalpy of a liquid stream will just reflect sensible heat changes. (OK – there are other enthalpy changes like heat of mixing, but we'll ignore these.) Sensible heat – energy change with temperature – is calculated using the heat capacity at constant pressure, C_P (kJ/mol °C),

$$C_P = a + bT + cT^2 + dT^3 \qquad (4.2)$$

and integrating over the temperature range from the reference state to the temperature of interest.

$$\hat{H} = \int_{T_R}^{T} C_P \, dT = a(T - T_R) + b(T^2 - T_R^2)/2 + c(T^3 - T_R^3)/3 + d(T^4 - T_R^4)/4 \quad (4.3)$$

In the heat capacity polynomial, a, b, c, and d are constants found in the appendix. Don't confuse the differential dT in the integral with the constant d in the fourth term. The heat capacity of liquids often has fewer terms than vapors.

The enthalpy of a vapor stream relative to a liquid reference condition involves both sensible and latent heat changes. The latent heat – energy associated with phase change – is usually known at a specific temperature such as the normal boiling point. This requires a hypothetical path calculation, summing the sensible heat to raise the liquid temperature from the reference temperature (T_R) to the boiling point (T_b), the latent heat to vaporize the liquid ($\hat{H}_V$), and the sensible heat to cool/heat the vapor to the final temperature (T).

$$\hat{H} = \int_{T_R}^{T_b} C_{PL} \, dT + \Delta \hat{H}_V + \int_{T_b}^{T} C_{PV} \, dT \qquad (4.4)$$

If a pressure change is involved, as with a flash drum, the enthalpy change for a liquid can be calculated as

$$\Delta \hat{H} = \Delta \hat{V} P = \hat{V} \Delta P \qquad (4.5)$$

for an incompressible liquid ($\Delta \hat{V} = 0$), where $\hat{V}$ is the specific volume (m³/mol) and ΔP is the change in pressure (atm). This enthalpy change is usually very small compared to latent and sensible heat changes. In fact, for an ideal gas the change in enthalpy with pressure at constant temperature is zero ($\Delta \hat{V} P = \Delta R T = 0$).

The total enthalpy of a stream will include the contribution from each species. For the two-component benzene–toluene system, a liquid stream enthalpy will be, for example,

$$H_L = L\left(x_B \hat{H}_B + x_T \hat{H}_T \right) \qquad (4.6)$$

where the units for the benzene term ($L x_B \hat{H}_B$) for example, are

$$\frac{\text{mol}}{s} \left| \frac{\text{mol B}}{\text{mol}} \right| \frac{kJ}{\text{mol B}} [=] \frac{kJ}{s}$$

Let's apply this to the same partial evaporator described in Example 3.1.

EXAMPLE 4.1 Partial Evaporator Energy Balance

A liquid feed of 100 mol/s comprising 40 mol% benzene and 60 mol% toluene is partially evaporated at 760 mm Hg. The vapor product flow rate is twice that of the liquid product flow rate. Determine all unknown flow rates and compositions. Also, calculate the required heat input.

   a. Draw a labeled process flow diagram,

   b. conduct a degree of freedom analysis, and

   c. write the equations and solve.

     In Example 3.1, a labeled process flow diagram (PFD) was constructed, a degree of freedom (DOF) analysis was made, and the temperature, flow rates, and compositions were determined. Recall that $V = 66.7$ mol/s, $L = 33.3$ mol/s, $x_B = 0.265$ mol B/mol, $y_B = 0.467$ mol B/mol, and $T = 99.7°C$ as determined by both mathematical and graphical methods. We can just pick up the problem at this stage. Only one unknown remains and that is the required heat, Q (kJ/s). It can be determined using the energy balance equation.

     We will need to define the thermodynamic state of the feed when calculating the energy required. Let's choose liquid at 760 mm Hg and 88.0°C since this is in the liquid-phase region (look at the Txy graph and/or check the bubble point equation $K_B x_B + K_T x_T < 1$). Let's choose the feed liquid as the reference state because it is at the lowest enthalpy state. That way, the enthalpy of this stream is conveniently and automatically zero and all other streams will have positive values. (Of course, you could choose any stream or even some arbitrary condition – the final answer will not be affected.) We define the specific enthalpy of benzene as $\hat{H}_B$ (liquid, 88°C, 760 mm Hg) $\equiv 0$ kJ/mol and likewise for the toluene in the feed stream.

     A labeled PFD is shown in Figure 4.1 with values for all process variables identified except Q. Just to be different, notice that the mole fractions of toluene are also labeled.

   c. Write the equations and solve.

     The evaporator increased the temperature and created a vapor phase requiring large amounts of energy added. To calculate the required heat, we need the enthalpy of all streams. Recall that we chose the feed stream conditions to be the reference state so that its enthalpy is defined as zero.

     For the liquid product, only the temperature is different, so the change in enthalpy is just the sensible heat required to

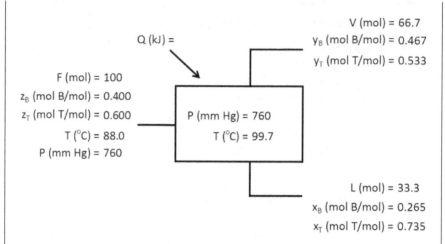

FIGURE 4.1
A labeled process flow diagram of a partial evaporator for use in Example 4.1.

increase the temperature from 88.0°C to 99.7°C. Heat capacity data is found in the appendix and for liquids it often has just two terms.

$$\hat{H}_i = \int_{88.0}^{99.7} C_{PL}\, dT$$

$$\hat{H}_B = 1.27 \times 10^{-1}(99.7 - 88.0) + 2.34 \times 10^{-4}\left(99.7^2 - 88.0^2\right)/2 = 1.73\ \text{kJ/mol}$$

$$\hat{H}_T = 1.49 \times 10^{-1}(99.7 - 88.0) + 3.24 \times 10^{-4}\left(99.7^2 - 88.0^2\right)/2 = 2.09\ \text{kJ/mol}$$

The total enthalpy of the liquid stream is then

$$H_L = L\left(x_B \hat{H}_B + x_T \hat{H}_T\right) = 33.3(0.265 \times 1.73 + 0.735 \times 2.09) = 66.6\ \text{kJ/s}$$

Calculating the enthalpy of the vapor is just a little more difficult because it involves a phase change. We usually don't know the heat of vaporization, $\Delta \hat{H}_V$, except at the normal boiling point. So we will employ a "hypothetical path" by changing the temperature of the liquid from the reference temperature to the boiling point, evaporating the liquid, and then changing the temperature of the vapor from the boiling point to the operating temperature.

$$\hat{H}_B = \int_{88.0}^{80.1} C_{PL}\, dT + \Delta\hat{H}_V + \int_{80.1}^{99.7} C_{PV}\, dT$$

$$= 1.27 \times 10^{-1}(80.1 - 88.0) + 2.34 \times 10^{-4}\left(80.1^2 - 88.0^2\right)\big/2 + 30.765$$

$$+ 7.41 \times 10^{-2}(99.7 - 80.1) + 3.30 \times 10^{-4}\left(99.7^2 - 80.1^2\right)\big/2$$

$$- 2.52 \times 10^{-7}\left(99.7^3 - 80.1^3\right)\big/3 + 7.76 \times 10^{-11}\left(99.7^4 - 80.1^4\right)\big/4$$

$$= 31.60 \text{ kJ/mol}$$

$$\hat{H}_T = \int_{88.0}^{110.6} C_{PL}\, dT + \Delta H_V + \int_{110.6}^{99.7} C_{PV}\, dT$$

$$= 1.49 \times 10^{-1}(110.6 - 88.0) + 3.24 \times 10^{-4}\left(110.6^2 - 88.0^2\right)\big/2 + 33.470$$

$$+ 9.42 \times 10^{-2}(99.7 - 110.6) + 3.80 \times 10^{-4}\left(99.7^2 - 110.6^2\right)\big/2$$

$$- 2.79 \times 10^{-7}\left(99.7^3 - 110.6^3\right)\big/3 + 8.03 \times 10^{-11}\left(99.7^4 - 110.6^4\right)\big/4$$

$$= 36.13 \text{ kJ/mol}$$

The total enthalpy of the vapor stream is

$$H_V = V\left(y_B\hat{H}_B + y_T\hat{H}_T\right) = 66.7(0.611 \times 31.60 + 0.389 \times 36.13) = 2{,}267.6 \text{ kJ/s}$$

Therefore, the heat required for this operation is

$$Q = H_L + H_V - 0 = 66.6 + 2{,}267.6 = 2{,}334 \text{ kJ/s}$$

Pencil in this value for heat added to Figure 4.1.

d. Explore the problem.

Go have a cup of coffee (it is important to take a short break) and come back to solve this problem on your own without looking at the solution. Solve for the enthalpies by hand and then check out Excel "Partial Evaporator/1-Stage." Try changing process conditions according to the suggested experiments or make up your own.

The conditions used in Example 4.1 were inputted to the flash simulator in ChemCAD. The thermodynamics wizard chose the non-random two-liquid (NRTL) model for K values and the LATE (latent heat) model for enthalpies. We might

TABLE 4.1

Comparison of Partial Evaporator Analysis using ChemCAD versus Hypothetical Paths in Example 4.1

Parameter	Hypothetical Path	ChemCAD
T (°C)	99.7	99.3
y_B (mol B/mol)	0.467	0.469
x_B (mol B/mol)	0.265	0.262
Q (kJ/s)	2,334	2,322

expect a slightly different solution but the results are close as shown in Table 4.1. Using Raoult's law in ChemCAD gave identical results to our calculations.

e. Be awesome!

If you've come this far, you are pretty awesome already! But if you want to be *REALLY AWESOME*, modify the spreadsheet to use liquid at the normal boiling point for the reference condition for each species. (This is the reference condition used by National Institute of Standards and Technology (NIST) to be discussed later in this chapter.) This means that no stream has zero enthalpy and the reference condition for each species is not the same. However, you'll find the final answer, the amount of required heat, is the same.

TRY THIS AT HOME 4.1   Partial Condenser Energy Balance

A vapor mixture of 100 mol/s comprising 55 mol% benzene and 45 mol% toluene is sent to a partial condenser operating at 760 mm Hg. 75 mol% of the toluene in the feed is recovered in the liquid product. Determine all unknown flow rates, compositions, and required heat removal.

a. Draw a labeled process flow diagram.

A labeled PFD is shown in Figure 4.2. Pencil in any information given in the process description. Leave the other variables blank for now.

Close the book right now and try the DOF analysis. You'll find two DOFs and need to specify the feed conditions to solve the problem. I'll complete parts (b) and (c) of our systematic procedure except you'll calculate the required heat.

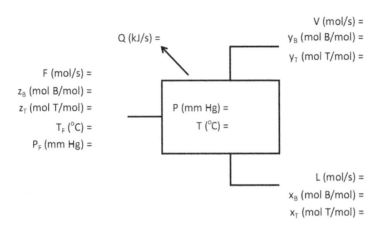

FIGURE 4.2
A labeled process flow diagram of a partial condenser for use in Try This at Home 4.1.

b. Conduct a degree of freedom analysis.

Unknowns (10): T_F, P_F, T, V, y_B, y_T, L, x_B, x_T, Q

Equations (8):

 2 mole balances (total and B)

 2 equilibrium relationships (Raoult's law for B and T)

 2 summations (vapor and liquid mole fractions must add to 1)

 1 process specification (75 mol% of toluene recovered)

 1 energy balance

There are 8 equations with 10 unknowns, or two DOFs. We can specify the temperature and pressure of the feed as $T_F = 102°C$ and $P_F = 760$ mm Hg. Take a moment to verify that this mixture is a vapor under these conditions by identifying the condition on the Txy diagram or checking the dew point equation ($y_B/K_B + y_T/K_T > 1$). Now there are zero DOF and the problem is solvable.

c. Write the equations and solve.

Total mole balance: $F = V + L$

Benzene mole balance: $Fz_B = Vy_B + Lx_B$

Benzene equilibrium: $y_B = K_B x_B$

Toluene equilibrium: $y_T = K_T x_T$

Vapor summation: $\sum y_i = 1$

Liquid summation: $\Sigma x_i = 1$

Process specification: $0.75\ Fz_T = L\ x_T$

If the heat is to be determined, the material balance equations above can be solved first and then the energy balance. Because the temperature is unknown, this iterative calculation is best done using Excel's Solver tool. The solution determined in Excel "Partial Condenser/1-Stage" is $V = 36.2$ mol/s, $y_B = 0.689$ mol B/mol, $y_T = 0.311$ mol T/mol, $L = 63.8$ mol/s, $x_B = 0.471$ mol B/mol, $x_T = 0.529$ mol T/mol, and $T = 93.0°C$. Verify by hand that these values satisfy the equations above.

Choose the lowest enthalpy stream for a reference condition. This is the liquid product and by our choice $\hat{H}_L \equiv 0$. Calculate the specific enthalpy of the benzene and toluene in both the vapor feed and the vapor product using heat capacities and heats of vaporization in the appendix. You can do this by hand since T is known (no iteration). Compare your solution with Excel "Partial Condenser/1-Stage."

$\hat{H}_B$ (feed vapor) =   Ans: 31.11 kJ/mol

$\hat{H}_T$ (feed vapor) =   Ans: 35.55 kJ/mol

$\hat{H}_B$ (product vapor) =   Ans: 30.17 kJ/mol

$\hat{H}_T$ (product vapor) =   Ans: 34.38 kJ/mol

Calculate the heat removed by the process using the energy balance equation.

$Q =$ Ans: −2,172 kJ/s

d. Explore the problem.

Using Excel "Partial Condenser/1-Stage," check the answers for this particular problem and then try the suggested experiments and other sets of starting values. Compare results with your study group. Take turns explaining each result.

e. Be awesome!

Make your own spreadsheet simulation from scratch using Excel "Partial Condenser/DIY." Check to make sure our answers are identical.

4.2 Trick of the Trade: The Enthalpy Table

To help organize your calculations, construct an enthalpy table. For Example 4.1: Partial Evaporator Energy Balance, it would look like this:

Reference Condition: Liquids at 760 mm Hg and 88.0°C.

Species	n_{in} (mol/s)	$\hat{H}_{in}$ (kJ/mol)	H_{in} (kJ/s)	n_{out} (mol/s)	$\hat{H}_{out}$ (kJ/mol)	H_{out} (kJ/s)
B (L)	50	0	0			
T (L)	50	0	0			
B (V)	---	---	---			
T (V)	---	---	---			
Total			0			

There is no vapor feed so n_{in}, $\hat{H}_{in}$, and H_{in} are not applicable for vapor and we enter "---". The liquid feed conditions were chosen as the reference state so the enthalpy is defined to be zero. Enthalpy calculations can be done right in the table (in Excel) by calling on data and physical constants from the PFD and other resources. Take a peek at the contents of cells in Excel "Partial Evaporator/1-Stage" to see how this was done.

TRY THIS AT HOME 4.2 Enthalpy Table for a Partial Condenser

Construct an enthalpy table (below) for Try This at Home 4.1. Compare your table to the one shown in Excel "Partial Condenser/1-Stage."

Species	n_{in} (mol/s)	$\hat{H}_{in}$ (kJ/mol)	H_{in} (kJ/s)	n_{out} (mol/s)	$\hat{H}_{out}$ (kJ/mol)	H_{out} (kJ/s)
B (L)						
T (L)						
B (V)						
T (V)						
Total						

4.3 Adiabatic Flash Drum

A vapor phase can be created by releasing a liquid into a vessel operating at a lower pressure and temperature. Q in this case is defined to be zero, necessitating that the material and energy balances be solved simultaneously.

Trust me – this type of trial and error calculation is much easier done by a computer. It takes Excel about a second to solve this problem and would take you and a calculator all day. One important difference from previous examples is that the enthalpy calculations will include a change in pressure using Eq. 4.5 in addition to temperature and phase changes.

EXAMPLE 4.2 Adiabatic Flash Drum

A total of 100 mol/s of an equimolar mixture of benzene and toluene at 133°C and 2,280 mm Hg (3 atm) is fed to an adiabatic flash drum operating at 760 mm Hg. (How do we know this feed mixture is a single liquid phase? Because the bubble point equation is $x_B K_B + x_T K_T < 1.0$ – try demonstrating this. You could also generate a Txy graph for 3 atm and see that this condition is below the bubble point curve – you are strongly encouraged to try this! Instructions for doing this are given in Excel "Adiabatic Flash/VLE.") Determine the temperature, flow rates, and compositions of the product vapor and liquid.

    a. Draw a labeled process flow diagram.
        The PFD is drawn in Figure 4.3. Unknown process variables are left blank. Don't forget to label Q (kJ/s) $\equiv 0$ (for an adiabatic process).
    b. Conduct a degree of freedom analysis.

    Unknowns (7): $T, V, y_B, y_T, L, x_B, x_T$

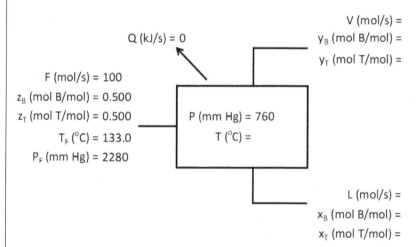

FIGURE 4.3
A labeled process flow diagram of an adiabatic flash drum for use in Example 4.2: Adiabatic Flash Drum.

Equations (7):

2 mole balances (total and B)

2 equilibrium relationships (Raoult's law for B and T)

2 summations (vapor and liquid mole fractions must add to 1)

1 energy balance

It looks like there are zero DOF and the problem can be solved.

"Wait a minute!" you might be saying. "What about enthalpies? Why haven't we included those as unknowns?" They will be used in the energy balance to force $Q = 0$. However, we'll consider those to be secondary calculations, just like we did with K values. If we know the temperature, pressure, and composition for each stream and have chosen a reference state, all enthalpies can be determined using a hypothetical path.

c. Write the equations and solve.

Since Q is specified, we have to solve the material and energy balance equations simultaneously.

Total mole balance: $F = V + L$

Benzene mole balance: $Fz_B = Vy_B + Lx_B$

Benzene equilibrium: $y_B = K_B x_B$ (remember that $K_i = P_i^*/P$)

Toluene equilibrium: $y_T = K_T x_T$

Energy balance: $Q = H_{out} - H_{in} = 0$

Liquid summation: $x_B + x_T = 1$

Vapor summation: $y_B + y_T = 1$

where F, V, L [=] mol/s and x, y, z [=] mol benzene (or toluene)/mol, and Q, H [=] kJ/s.

It is essential to set up the enthalpy table to organize those calculations. Let's choose the feed stream for a reference state since we know its complete thermodynamic state, although we could easily use any reference state.

Reference Condition: Liquids at 2,280 mm Hg and 133.0°C.

Species	n_{in} (mol)	$\hat{H}_{in}$ (kJ/mol)	H_{in} (kJ)	n_{out} (mol)	$\hat{H}_{out}$ (kJ/mol)	H_{out} (kJ)
B (L)	50	0	0			
T (L)	50	0	0			
B (V)	---	---	---			
T (V)	---	---	---			
Total						

Since we don't know temperature, this is a highly iterative solution. Use Excel "Adiabatic Flash/1-Stage." Input initial guesses into the simulation and then let the Solver tool iterate until all equations are satisfied. We can use the Txy graph to choose reasonable initial *guesses* for this two-phase system: $T = 96°C$, $x_B = 0.45$ mol B/mol, $y_B = 0.55$ mol B/mol, $L = V = 50$ mol/s.

Calculation of Enthalpies

To calculate the enthalpy of the liquid product, we employ a hypothetical path consisting of a change in pressure from 3 atm to 1 atm, followed by a change in temperature from 133.0°C to the (unknown) operating temperature. The equation for benzene is

$$\hat{H}_{B,L} = \hat{V}\Delta P + \int_{133}^{T} C_{PL}\, dT \tag{4.7}$$

The first term for change in pressure included an assumption that the liquid is incompressible (constant $\hat{V}$). Calculation involves some unit conversions including the handy 100 J/L bar. Be sure to use the "picket fence" method for unit clarity.

$$V\Delta P = \frac{78.11\text{ g}}{\text{mol}} \left| \frac{\text{L}}{879\text{ g}} \right| \frac{(760 - 2{,}280)\text{ mm Hg}}{} \left| \frac{1\text{ atm}}{760\text{ mm Hg}} \right.$$

$$\frac{1.01325\text{ bar}}{\text{atm}} \left| \frac{100\text{ J}}{\text{L bar}} \right| \frac{\text{kJ}}{1{,}000\text{ J}}$$

$$= -0.018\text{ kJ/mol}$$

As an exercise, cancel out the units in the above equation. The second term can only be calculated with a guessed operating temperature, say 96°C.

$$\int_{133}^{96} C_P\, dT = 1.27 \times 10^{-1}(96 - 133) + 2.34 \times 10^{-4}\left(96^2 - 133^2\right)/2 = -5.67\frac{\text{kJ}}{\text{mol}}$$

Both terms are negative because the product is at a lower temperature and lower pressure. The specific enthalpy of benzene in the liquid product is then

$$\hat{H}_{B,L} = -0.02 - 5.67 = -5.69 \frac{kJ}{mol}$$

The product vapor enthalpy also includes the change in pressure, followed by change in liquid feed temperature to the normal boiling point, vaporization, and finally change in vapor temperature to the guessed operating temperature. (Try this calculation yourself!)

$$\hat{H}_{B,V} = \hat{V}\Delta P + \int_{133}^{80.1} C_{PL}\, dT + \Delta\hat{H}_V + \int_{80.1}^{96} C_{PV}\, dT = 24.34\ kJ/mol$$

Although Excel "Adiabatic Flash/Flash Drum" includes calculating the VΔP term, it is very small compared to the sensible and latent heat terms and could be ignored. In fact, the final split of liquid and vapor is mostly a trade-off of sensible and latent heat terms. As an exercise, calculate the specific enthalpy of the toluene liquid and vapor products.

$\hat{H}_{T,L} =$ Ans: −6.90 kJ/mol

$\hat{H}_{T,V} =$ Ans: 27.33 kJ/mol

As Solver iterates, it chooses different flash temperatures that, in turn, generate new compositions and enthalpies. Using an efficient algorithm, it converges on a solution to all of our equations (including Q = 0). The final answer is T = 93.5°C, L = 78.7 mol/s, $x_B = 0.453$ mol B/mol, $x_T = 0.547$ mol T/mol, V = 21.3 mol/s, $y_B = 0.673$ mol B/mol, and $y_T = 0.327$ mol T/mol. The completed enthalpy table is shown below (Table 4.2).

TABLE 4.2

Enthalpy Table in Excel "Adiabatic Flash Drum"

Reference Condition: Liquids at Feed Conditions

Species	n_{in} (mol)	H_{in} (kJ/mol)	H_{in} (kJ)	n_{out} (mol)	H_{out} (kJ/mol)	H_{out} (kJ)
B (L)	50.0	0.0	0.0	35.7	−6.1	−216.3
T (L)	50.0	0.0	0.0	43.0	−7.3	−316.3
B (V)	---	---	---	14.3	24.1	344.7
T (V)	---	---	---	7.0	27.0	188.0
Total			0.0			0.0

TABLE 4.3

Enthalpies Calculated in Excel "Adiabatic Flash Drum" for Each Step
in the Hypothetical Path

	Enthalpy Calculations From Reference State				
Prod. Liquid	**B**	**T**	**Prod. Vapor**	**B**	**T**
Ref State	0	0	Ref State	0	0
ΔP	−0.02	−0.02	ΔP	−0.02	−0.02
$T_F - T$	−6.04	−7.33	$T_F - T_B$	−8.01	−4.21
			ΔH_V	30.77	33.47
			$T_B - T$	1.35	−2.23
H (kJ/mol)	−6.06	−7.35		24.09	27.01

The final enthalpy calculations for each step in the hypothetical paths are shown in Table 4.3.

d. Explore the problem.

Check out Excel "Adiabatic Flash/1-Stage." Make sure you
understand the calculations in each cell before continuing. The
first thing you should do is repeat the above example to make
sure you get the same results. Perform all calculations by hand
to verify the results and to practice your math skills.

Now try various feed and operating conditions. What happens if your feed temperature or pressure is different? With
a small modification of Solver, you can specify a temperature
and calculate the heat required. Try the suggested experiments
listed in the spreadsheet.

e. Be awesome!

Make your own simulation by using Excel "Adiabatic Flash/
DIY" and inserting all of the relevant equations. After running
Solver, make sure our answers are the same.

The conditions used in Example 4.2 were inputted to the
flash simulator in ChemCAD. The thermodynamics wizard
chose the NRTL model for K values and the LATE (latent heat)
model for enthalpies. We might expect a slightly different solution but the results are close as shown in Table 4.4.

4.4 Using Published Thermodynamic Data

We have calculated the enthalpy of benzene/toluene mixtures relative to a
reference state of our choosing with a combination of sensible and latent heat

TABLE 4.4

Comparison of Adiabatic Flash Drum Analysis using ChemCAD versus Hypothetical Paths

Parameter	Hypothetical Path	ChemCAD
T (°C)	93.5	93.2
V (mol)	21.3	21.9
y_B (mol B/mol)	0.673	0.670
L (mol)	78.7	78.1
x_B (mol B/mol)	0.453	0.452

changes along a hypothetical path. We can also use thermodynamic data from a variety of open and commercial sources. A convenient resource for a variety of chemicals is provided by the NIST. The website is http://webbook. nist.gov/chemistry/fluid/.

Use NIST data to calculate the heat required for Example 4.1. Open the NIST web page and follow these steps: choose (1) Benzene, (2) Celsius, atm, kJ/mol (the other units don't matter), (3) isobaric properties, (4) default, and continue. On the next page, choose (1) 1 atm, (2) 76°C to 120°C by 2°C increments, and press for data. The first plot that comes up is density versus temperature. Change the y-axis to enthalpy and you will see it is mostly vapor in this range because the normal boiling point is 80.084°C. Under "other data available" choose "Download data as a tab-delimited text file." Copy and paste into Excel as a text file. You can format this data as you see fit. Repeat these steps for toluene. For now, look at *Excel "NIST"* that also includes well-formatted enthalpy data for both benzene and toluene.

Notice that the enthalpy is 0 kJ/mol for liquid benzene at 80.1°C and for liquid toluene at 110.6°C. These temperatures are the normal boiling points. You now recognize that the reference condition must be liquid at the normal boiling point. It is perfectly fine to use different reference conditions for each chemical species in an energy balance calculation as long as there is no chemical reaction. The data is for benzene vapor above 80.1°C and for toluene liquid below 110.6°C because these results assume a single component. But we know that between the normal boiling points we can have a two-phase mixture as shown on the Txy graph. In order to estimate the enthalpy of liquid benzene above 80.1°C or vapor toluene below 110.6°C we need to assume the heat of vaporization, $\Delta \hat{H}_V$ (kJ/mol), is independent of temperature. Then just add $\Delta \hat{H}_V$ to liquid enthalpies and subtract it from vapor enthalpies. Study Excel "NIST/H Data" to see how this was done.

This data is plotted in Excel "NIST/B H Chart" and "NIST/T H Chart" and reproduced in Figures 4.4 and 4.5. The data is fitted to a straight-line equation that provides a simple way to calculate both liquid and vapor enthalpies

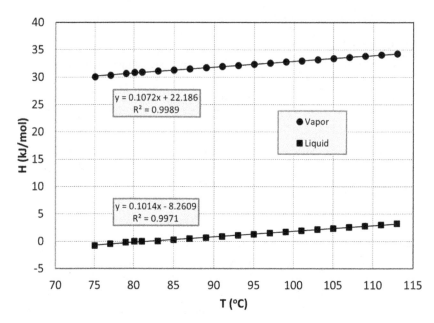

FIGURE 4.4
Liquid and vapor enthalpies of benzene at 1 atm from NIST.

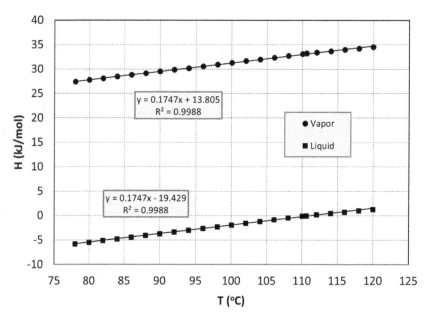

FIGURE 4.5
Liquid and vapor enthalpies of toluene at 1 atm from NIST.

as a function of temperature. Now we can calculate the heat required in the partial evaporator of Example 4.1 using these enthalpies. Just remember that we now have a new reference condition (liquids, 1 atm, at their normal boiling point) and not the condition of the feed stream.

EXAMPLE 4.3 Partial Evaporator Energy Balance Using NIST Data

Recall that 100 mol/s of a liquid mixture with 40 mol% benzene and 60 mol% toluene was fed to a partial evaporator operating at 760 mm Hg in Example 4.1. The vessel was heated to produce twice as much vapor as liquid product. Our job was to determine the evaporator's temperature and the amount and composition of each phase. We used a combination of latent and sensible heat changes to calculate enthalpies relative to the feed state and found that the required heat input was 2,334 kJ/s. Now we will use the NIST data to determine the required heat input and compare.

a. Draw a labeled PFD,

b. Conduct a DOF analysis, and

c. Write the equations and solve.

    All this was done in Example 3.1: Partial Evaporator Analytical Solution and copied in Example 4.1: Partial Evaporator Energy Balance. The material balance equations can be solved first and then the energy balance to determine the heat input. The flow rates, compositions, and temperature will be the same, of course.

    The point of this example is to repeat the enthalpy calculations using the NIST data. The enthalpies of benzene liquid feed and toluene product vapor are calculated from the straight-line equations.

$$\hat{H}_{B,L}(\text{feed}) = 0.1014 \times 88°C - 8.2609 = 0.66 \text{ kJ/mol}$$

$$\hat{H}_{T,V}(\text{product}) = 0.1747 \times 99.7°C + 13.805 = 31.22 \text{ kJ/mol}$$

These are shown in Table 4.5 reproduced from Excel "NIST/Partial Evaporator." Use the straight-line equations in Figures 4.4 and 4.5 to verify the other specific enthalpies.

    The energy balance is

$$Q = H_{\text{out}} - H_{\text{in}} = 2,100 - (-217) = 2,317 \text{ kJ/s}$$

TABLE 4.5

Enthalpy Table for Partial Evaporator Using Data in Excel "NIST"

Reference Condition: Liquids at Normal Boiling Point

Species	n_{in} (mol)	H_{in} (kJ/mol)	H_{in} (kJ)	n_{out} (mol)	H_{out} (kJ/mol)	H_{out} (kJ)
B (L)	40.00	0.66	26.49	8.84	1.85	16.34
T (L)	60.00	−4.06	−243.32	24.49	−2.01	−49.32
B (V)	---	---	---	31.16	32.87	1,024.15
T (V)	---	---	---	35.51	31.22	1,108.65
Total			−216.83			2,099.81

This is very close to the 2,334 kJ/s calculated using hypothetical paths and the feed stream as a reference condition.

d. Explore the problem.

Repeat any single-stage energy balance problem that you previously solved and compare the answers.

e. Be awesome!

Now that you are getting more comfortable using these Excel simulations (hopefully!), modify the partial condenser and adiabatic flash simulations to use these NIST enthalpies and compare the results to calculations using hypothetical paths for enthalpies. Note that for the adiabatic flash we didn't find the enthalpy at 3 atm. However, the effect of dropping the pressure on enthalpy was negligible and we are relatively safe to ignore that.

4.5 Summary

We added the required heat transfer, Q, as an unknown in analyzing a partial evaporator, a partial condenser, and an adiabatic flash drum. This generally requires adding one more equation, the energy balance. If Q is to be calculated, we can solve the material balances first and then the energy balance. If it is specified (for instance, zero in an adiabatic process) the material and energy balance equations must be solved simultaneously. The energy balance requires calculating enthalpies and this was done using hypothetical paths, NIST data, and the ChemCAD chemical process simulator.

5

Multi-Stage Distillation

You may have noticed that the amount of separation in a single equilibrium stage was limited. The operating lines on an xy graph show this quite clearly. Figure 5.1 shows the equilibrium vapor and liquid mole fractions of benzene with toluene at 1 atm. The intersection of the operating line and the 45° line is the feed composition – in the case of an equimolar mixture with 0.50 mol B/mol. If only one bubble of vapor is produced, virtually all liquid product, then the slope of the operating line is $-L/V \sim \infty$ (see Eq. 3.7), resulting in the vertical line. This yields the highest benzene vapor concentration possible (0.71 mol B/mol) but with an impracticably small amount of vapor product. If only one drop of liquid is produced, virtually all vapor product, then the slope of the operating line is $-L/V \sim 0$, resulting in the horizontal line. This yields the lowest benzene liquid concentration possible (0.29 mol B/mol) but with an impracticably small amount of liquid product. One of the suggested experiments in Excel "Single Stage/1-Stage" was to examine these limits. Go ahead and try this if you haven't already. It is

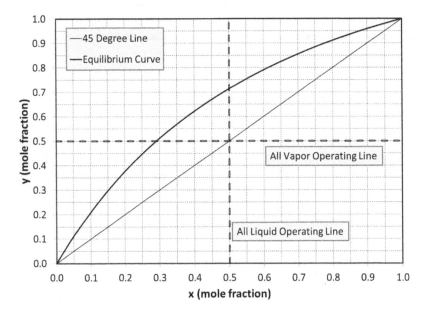

FIGURE 5.1
Vapor-liquid equilibrium (VLE) of benzene and toluene at 1 atm. Operating lines for all vapor and all liquid show the possible range of benzene composition in the products.

difficult to obtain good separation of chemicals in a single equilibrium stage unless the relative volatilities are very different making the equilibrium line far away from the 45° line.

5.1 Analysis of Two Stages

Much better separation can be achieved by connecting multiple equilibrium stages in series. This is the basic concept of distillation, the most common type of separation process, and that can commonly consist of 20 or more stages. We'll start by looking at two stages in series, which produces significantly purer products. Then we'll connect multiple stages below and above the feed stage to get even better separated top (benzene-rich) and bottom (toluene-rich) products. Energy balances will be postponed until Chapter 8.

EQUILIBRIUM STAGE OR TRAY?

I will usually refer to equilibrium stages, or just stages, rather than trays in a distillation column. There is not a one-to-one correspondence because trays are usually not 100% efficient. More trays will be needed than your count of equilibrium stages indicate. No worries – we'll learn how to account for this.

EXAMPLE 5.1 Two Stages in Series, No Recycle

A total of 100 mol/s of a liquid mixture with 40 mol% benzene and 60 mol% toluene is fed to the bottom stage (numbered from the top so this is Stage 2) of a two-stage separator operating at 1 atm. Heat is added to Stage 2 to vaporize 2/3 of the feed. The vapor product then flows to Stage 1, where heat is removed to condense 1/3 of the vapor fed to it. Our job is to determine the flow rate and composition of each stream.

 a. Draw a labeled process flow diagram.
    From the process description, we know the feed flow rate, composition, and pressure. A process flow diagram (PFD) with these values is shown in Figure 5.2 and the remaining variables are unknown. Flow rates and compositions of the streams are numbered according to the stage from which they leave.

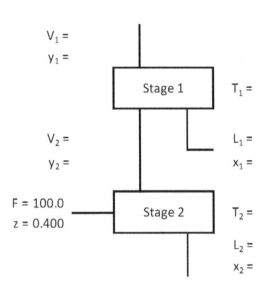

$V_1 =$

$y_1 =$

Stage 1 $T_1 =$

$V_2 =$ $L_1 =$

$y_2 =$ $x_1 =$

$F = 100.0$

$z = 0.400$ Stage 2 $T_2 =$

$L_2 =$

$x_2 =$

FIGURE 5.2

Two-stage partial evaporator with no liquid recycle operating at 1 atm. Feed, liquid, and vapor streams are labeled with flow rates (F, V, L [=] mol/s), compositions (x, y, z [=] mol B/mol), and temperature (T [=] °C).

b. Conduct a degree of freedom analysis.

For multi-unit processes, we first look at breaking the problem into solvable subsystems: overall process, Stage 1, or Stage 2. First, draw a material balance envelope (MBE) around Stage 2 in Figure 5.2. Consider only the streams crossing the MBE.

Unknowns (5): L_2, x_2, V_2, y_2, and T_2

Equations (5):

   2 mole balances (total and B)

   2 equilibrium relationships (Raoult's law for B and T)

   1 process specification (2/3 of feed vaporized)

There are zero degrees of freedom (DOF) so the unknowns crossing the Stage 2 MBE can be determined. In fact, this is the same problem as the single stage in Example 3.1. We found $V = 66.7$ mol/s, $L = 33.3$ mol/s, $x_B = 0.265$ mol B/mol, $y_B = 0.467$ mol B/mol, and $T = 99.7$°C. Pencil these values into Figure 5.2.

Now that V_2 and y_2 have been determined, draw an MBE around Stage 1 and conduct the DOF analysis.

Unknowns (5): V_1, y_1, L_1, x_1, and T_1

Equations (5):

   2 mole balances (total and B)

   2 equilibrium relationships (Raoult's law for B and T)

   1 process specification (1/3 of vapor condensed)

There are zero DOF for Stage 1 and the remaining unknowns can be determined.

c. Write the equations and solve.

   Here we find a small limitation of Excel's Solver tool. For a complete solution, we would need to make this calculation in two steps (Stage 2 and then Stage 1) requiring an annoying reformulation of Solver with each step. This would be particularly tedious if we were conducting many experiments.

   Instead, we can solve for all unknowns at once. The entire process has 10 unknowns (V, L, x, y, and T for each stage) and 10 equations (2 mole balances and 2 equilibrium relationships for each stage, and 2 process specifications for product splits). This yields zero DOF and the problem can be solved. This "brute force" method is easy enough to program in Excel's Solver tool. Looking forward, it is not all that hard to do with even more stages.

   The equations are:

Stage 1

Total mole balance: $V_2 = V_1 + L_1$

Benzene mole balance: $y_2 V_2 = y_1 V_1 + x_1 L_1$

Benzene equilibrium: $y_1 = K_{1,B} x_1$

Toluene equilibrium: $(1 - y_1) = K_{1,T}(1 - x_1)$

PS Stage 1: $V_1 = 2L_1$

Stage 2

Total mole balance: $F = V_2 + L_2$

Benzene mole balance: $Fz = y_2 V_2 + x_2 L_2$

Benzene equilibrium: $y_2 = K_{2,B} x_2$

Toluene equilibrium: $(1 - y_2) = K_{2,T}(1 - x_2)$

PS Stage 2: $V_2 = 2L_2$

The solution is shown graphically in Figures 5.3 and 5.4 with the operating lines for both stages.

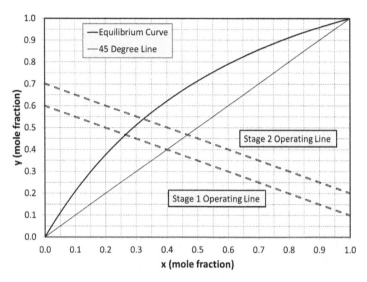

FIGURE 5.3
Benzene–toluene VLE at 1 atm showing the operating lines for the two-stage partial evaporator with no recycle.

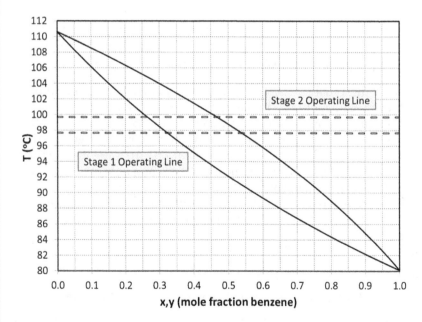

FIGURE 5.4
Benzene–toluene VLE at 1 atm showing the operating lines for the two-stage partial evaporator with no recycle.

The solution for Stage 2 is the same as in Example 3.1 because the specifications are identical. By adding Stage 1, we boost the vapor-phase composition from 0.467 for a single stage to 0.539 mol B/mol for these two stages. Unfortunately, this comes at the cost of reducing the amount of product by a third. Study the solution to this problem in Excel "Multi-Stage Distillation/2-Stage No Recycle." Pencil all remaining values into Figure 5.2.

Heat will need to be added to Stage 2 and removed from Stage 1. In a full multi-stage distillation column heat is added to a partial evaporator (reboiler) at the bottom of the column and removed with a partial condenser at the top of the column. Let's tuck that qualitative knowledge away for now and quantify it in Chapter 8.

d. Explore the problem.

Try the experiments suggested in Excel "Multi-Stage Distillation/2-Stage No Recycle." Can you optimize the process for different objectives?

e. Be awesome!

Make a copy of Excel "Multi-Stage Distillation/2-Stage No Recycle." Modify it to be a partial condenser with a vapor feed to Stage 1.

TRY THIS AT HOME 5.1   Fun with Two-Stage Separators

Solve this problem graphically. Print an xy graph from Excel "Multi-Stage Distillation/xy." Locate the operating line for Stage 2 using the composition of the feed as a point on the 45° line and L_2/V_2 for a slope. Using the composition of the vapor leaving Stage 2 as the feed to Stage 1 (hint: draw a horizontal line from the intersection point), locate the operating line for Stage 1, and note the final compositions. Compare your graphical flow rates and compositions to those obtained analytically and to the operating lines in Figures 5.3 and 5.4. Name this "new" graphical technique after yourself – the _____ method!

EXAMPLE 5.2   Two Stages in Series with Recycle

Consider the process described above in Example 5.1. The main difference is that we will recycle the liquid from Stage 1 back to Stage 2. The returned liquid is called reflux. This will avoid the problem of losing a third of our vapor product. We will still have two mole balances and two equilibrium equations for each stage. Now we will make the overall vapor product molar flow rate equal to twice the overall liquid product molar flow rate ($V_1 = 2L_2$) as one process specification. The other will be to specify the reflux stream molar flow rate to be the same as the molar flow rate of the vapor product ($L_1 = V_1$), a reflux ratio of 1.

a. Draw a labeled process flow diagram.

   The PFD is drawn in Figure 5.5. It looks identical to Figure 5.2 except that the liquid leaving Stage 1 is fed back to Stage 2. It looks like all flow rates, compositions, and temperatures are known or identified.

b. Conduct a degree of freedom analysis.

   Unknowns (10): V_1, y_1, T_1, V_2, y_2, L_1, x_1, T_2, L_2, x_2

   Equations (10):

   4 mole balances (total and B for each stage)

   4 equilibrium relationships (Raoult's law for B and T for each stage)

   2 process specifications (overall product split and reflux ratio)

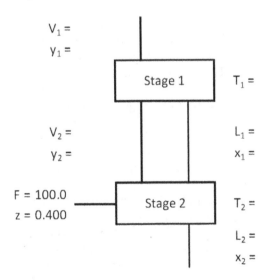

FIGURE 5.5

Two-stage partial evaporator with liquid recycle (reflux) operating at 1 atm. Feed, liquid, and vapor streams are labeled with flow rates (F, V, L [=] mol/s), compositions (x, y, z [=] mol B/mol), and temperature (T [=] °C).

The unknowns are identical to Example 5.1. The mole balances for Stage 2 must also include the reflux stream as an input. Also, the process specifications are different. This problem seems to be properly specified.

c. Write the equations and solve.
    The equations are:

Stage 1

Total mole balance: $V_2 = V_1 + L_1$

Benzene mole balance: $y_2V_2 = y_1V_1 + x_1L_1$

Benzene equilibrium: $y_1 = K_{1,B}x_1$

Toluene equilibrium: $(1 - y_1) = K_{1,T}(1 - x_1)$

Stage 2

Total mole balance: $F + L_1 = V_2 + L_2$ (Note the extra input stream L_1.)

Benzene mole balance: $Fz + L_1x_1 = y_2V_2 + x_2L_2$

Benzene equilibrium: $y_2 = K_{2,B}x_2$

Toluene equilibrium: $(1 - y_2) = K_{2,T}(1 - x_2)$

PS Overall flow rates: $V_1 = 2L_2$

PS Reflux: $L_1 = V_1$

Go ahead and solve these by hand if you like and we'll get back together next week. Or you can join me in using Excel "Multi-Stage Distillation/2-Stage" to solve these in about 1 s using Solver. The vapor stream from Stage 1 is $V_1 = 66.7$ mol/s with a composition of $y_1 = 0.495$ mol B/mol. Compare this to $y = 0.467$ mol B/mol for a single stage. The liquid stream from Stage 2 is $L_2 = 33.3$ mol/s with a composition of $x_2 = 0.210$ mol B/mol compared to $x = 0.265$ mol B/mol for a single stage. The vapor product is richer in benzene and the liquid product is richer in toluene by adding a second stage while the amounts of products are the same. Pencil these values into Figure 5.5 and all remaining unknowns from Excel "Multi-Stage Distillation/2-Stage."

Operating lines are not very useful for this situation. The intersection with the 45° line for Stage 2 is not at the feed composition 0.400, rather it is at the average concentration of the feed and reflux streams, which we don't know at first. In fact, it is very interesting that the vapor composition leaving Stage 2 ($y_2 = 0.391$) is actually lower (!) than the feed ($z = 0.400$) because of the diluting effect of the reflux stream. We can't use your graphical solution method here.

TABLE 5.1

Effect of Product Flow Rate and Reflux Ratio

Process		y_1
One Stage	V = 66.7	0.467
	V = 50.0	
	V = 33.3	
Two Stage	$V_1 = 66.7, R = 0$	0.495
	$V_1 = 66.7, R = 1$	
	$V_1 = 66.7, R = 2$	
	$V_1 = 50.0, R = 0$	
	$V_1 = 50.0, R = 1$	
	$V_1 = 50.0, R = 2$	
	$V_1 = 33.3, R = 0$	
	$V_1 = 33.3, R = 1$	
	$V_1 = 33.3, R = 2$	

Here is a qualitative question for you. Will the addition of recycle change the heating and cooling requirements for the process? Give that some thought and try to stump your study group, TA, or professor!

d. Explore the problem.

What do you think will happen if you change the amount of reflux or vapor product? Give it a try using Excel "Multi-Stage Distillation/2-Stage" and then pencil in your results into Table 5.1 and explain.

e. Be awesome!

Make a copy of Excel "Multi-Stage Distillation/2-Stage." Modify it to be a partial condenser with a vapor feed to Stage 1.

**TRY THIS AT HOME 5.2   More Fun
with Two-Stage Separators**

We will usually ignore this, but there is always some pressure drop going up the column. Think of several inches of the liquid in each stage in a vertical tray column exerting hydraulic pressure on the stages below. The spreadsheet assumes a pressure of 760 mm Hg but it doesn't appear explicitly. Where is it? Make a WAG (a technical engineering term for a guess!) of the expected pressure drop. Modify the spreadsheet to allow a lower pressure in Stage 1. Does it make much difference?

5.2 Analysis of Several Equilibrium Stages in Series

Better separation, by which I mean higher benzene concentration in the vapor product, can be achieved by taking less vapor product, increasing the reflux ratio, or adding more stages. We've seen how to analyze a two-stage system with the extra stage above the feed stage. Now we'll add two stages below and two above the feed stage for a simple five-stage distillation column.

Consider the PFD shown in Figure 5.6. A liquid mixture is fed to the middle, Stage 3. If it is fed at the bubble point, it will join the liquid flowing down the column (approximately – it depends on the thermodynamic state of the other streams entering the stage as we will see in Chapter 8). Heat is added to Stage 5 to partially vaporize the liquid from Stage 4. Heat is removed from Stage 1 to partially condense the vapor from Stage 2. Vapor is removed from the top stage and will be rich in the light component. Liquid is removed from the bottom stage and will be rich in the heavy component.

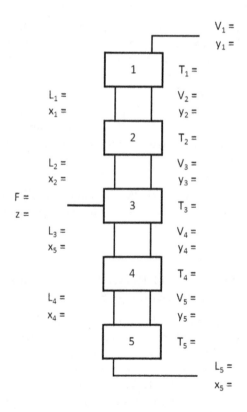

FIGURE 5.6
Five-stage distillation process operating at constant pressure. Feed, liquid, and vapor streams are labeled with flow rates (F, V, L [=] mol/s), compositions (x, y, z [=] mol B/mol), and temperature (T [=] °C).

The DOF analysis starts to get a bit complicated here, but let's give it a try. Let's say that the feed flow rate and composition, liquid product flow rate, and pressure were given. That was enough for a single stage. We could try breaking the system into smaller MBEs, for example, just around Stage 1. But no subsystem can be found with zero DOF (give this a try!). In Chapter 6, we'll find a more elegant solution to this problem, but for now let's just use the brute force method on the entire system.

The feed (F, z, T_F, P_F) is usually specified. That leaves five liquid and five vapor streams with their corresponding flow rates and compositions. Also, we have five temperatures for a total of 25 unknowns. There will be two mole balances and two equilibrium relationships for each stage. As before, we will include the reflux ratio (V_1/L_1) and one product molar flow rate (say L_5) as process specifications. This results in 22 algebraic equations, so the problem appears underspecified by three degrees.

We also need to make one other very important assumption: that the total molar flow rates of vapor or liquid in the column are constant, even though composition is changing, except where material is added (feed stage) or removed (top and bottom products). We will use this assumption to state that $V_2 = V_3$, $V_3 = V_4$, and $V_4 = V_5$, giving three more process specifications and zero DOF.

This is called the equimolal overflow (EMO) assumption and is based on thermodynamic considerations (also called constant molal overflow (CMO)). As the light component moves from the liquid to the vapor with the temperature decreasing going up the column, and the heavy component moves from the vapor to the liquid with the temperature increasing going down the column, these latent and sensible heat changes are assumed to be equal allowing the flow rates from each stage to be unchanged, except where material is added or removed. We will relax this assumption in Chapter 8 when we add enthalpy balances to the material balance equations.

EXAMPLE 5.3   Distillation in a Five-Stage Column

A total of 100 mol/s of a liquid mixture with 40 mol% benzene and 60 mol% toluene at the bubble point temperature is fed to the middle stage of a five-stage column operating at 760 mm Hg. The molar flow rate of the liquid product will be 33.3 mol/s and the reflux ratio, R, will be 1.0. EMO will be assumed. Determine all flow rates, compositions, and temperatures in the column.

a. Draw a labeled process flow diagram.
   We will use the PFD in Figure 5.6 with $F = 100$ mol/s, $z = 0.40$ mol B/mol, and $L_5 = 33.3$ mol/s. Please pencil in these values on the PFD. All other flow rates, compositions, and temperatures are left as unknowns.

b. Conduct a degree of freedom analysis.

Considering the entire system (brute force method):

Unknowns (24): L_{1-4}, x_{1-5}, V_{1-5}, y_{1-5}, T_{1-5}

Note that the difference here from the general discussion is that L_5 is specified, one less unknown.

Equations (24):

10 mole balances (total and B for each stage)

10 equilibrium relationships (Raoult's law for B and T for each stage)

4 process specification (reflux ratio, EMO for three stages)

The problem is properly specified. This is not as obvious as the two-stage systems. You should spend some extra time examining the DOF analysis.

c. Write the equations and solve.

For each of the five stages:

Total mole balance: $V_{i+1} + L_{i-1} = L_i + V_i$

Benzene mole balance: $V_{i+1}y_{i+1} + L_{i-1}x_{i-1} = V_iy_i + L_ix_i$

Note: For the feed stage, add F and Fz to the total and benzene mole balances. Also, there is no vapor flow to Stage 5 and no liquid flow to Stage 1.

Benzene equilibrium: $y_{i,B} = K_{i,B}x_{i,B}$

Toluene equilibrium: $(1 - y_{i,B}) = K_{i,T}(1 - x_{i,B})$

Also

PS (Stage 1) reflux ratio: $L_1 = RV_1$

PS (Stages 2–4) EMO: $V_i = V_{i+1}$

The solution appears in Excel "Multi-Stage Distillation/5-Stage." Pencil all values into the PFD in Figure 5.6. Now for the

TABLE 5.2

Effect of Multiple Stages in Series

Process	y (Vapor Product)	x (Liquid Product)
One Stage	0.467	0.265
Two Stage	0.495	0.210
Five Stage	0.566	0.067

TABLE 5.3

Effect of Liquid Product Flow Rate and Reflux Ratio for Five Stages in Series

Experiment	y (Vapor Product)	x (Liquid Product)
$R = 1, L_5 = 33.3$	0.566	0.067
$R = 1, L_5 = 50.0$		
$R = 1, L_5 = 66.7$		
$R = 2, L_5 = 33.3$		
$R = 0.5, L_5 = 33.3$		

same process conditions as in Examples 3.1 and 5.2 but with five stages, we achieve 0.566 mol B/mol while taking 66.7 mol/s of vapor product. The toluene purity increases in the liquid product also. The performance for this split of the products is summarized in Table 5.2 for one-, two-, and five-stage processes.

d. Explore the problem.

You already demonstrated using two stages, that increasing the reflux ratio or decreasing the amount of vapor product increases the mole fraction of benzene in the vapor product. Use Excel "Multi-Stage Distillation/5-Stage" to experiment with these variables. Try increasing the liquid product (L_5) or increasing the reflux ratio and record the results in Table 5.3. You can even try to make alternative specifications, such as the amount of vapor product or the ratio V_5/L_5 (similar to reflux ratio and called boilup.) A word of caution: if Solver can't find a solution, it may because you are asking this five-stage column to do the impossible. We'll get better at diagnosing that later.

e. Be awesome!

Modify a copy of Excel "Multi-Stage Distillation/5-Stage" to include seven stages with feed going to the middle stage #4. You should observe a significant improvement in separation.

TRY THIS AT HOME 5.3 An Alternate Feed Specification

It turns out that Stage 3 is not the optimum location for the feed in Example 5.3 "Distillation in a 5-Stage Column." Instead try F = 100 mol/s, z = 0.50 mol B/mol (still fed to Stage 3), $V_1 = L_5$, R = 1, and the three EMO constraints as specifications.

a. Draw a labeled process flow diagram,

b. Conduct a degree of freedom analysis, and

c. Write the equations and solve.

These are basically the same steps as Example 5.3. We have one more unknown (L_5) and one more process specification ($V_1 = L_5$, that is equal split of vapor and liquid products). Modify Excel "Multi-Stage Distillation/5-Stage" to solve this problem. You should find that $y_1 = 0.799$ mol B/mol and $x_5 = 0.201$ mol B/mol.

d. Explore the problem.

Here is a very interesting observation. The composition of the passing streams (x_i, y_{i+1}) (dots) are related by material balances and the leaving streams (x_i, y_i) (squares) are related by equilibrium. They are plotted in Figure 5.7. This plot is also found in Excel "Multi-Stage Distillation/5-Stage Streams" and updated with every trial. Note that the top point is (y_1, y_1) and the bottom point is (x_5, x_5) since there are no passing liquid and vapor streams at Stages 1 and 5, respectively.

If you take a pencil and ruler to draw a line through the top three and bottom three passing stream compositions,

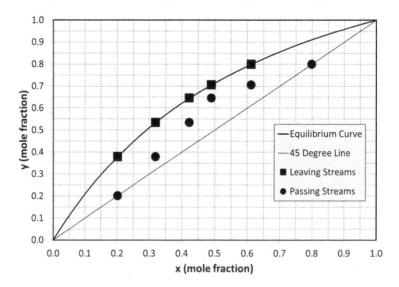

FIGURE 5.7

Benzene–toluene VLE at 1 atm showing the stream compositions for the five-stage distillation process. The squares represent composition of liquid and vapor streams leaving a stage in equilibrium. The circles represent composition of liquid and vapor streams passing between stages.

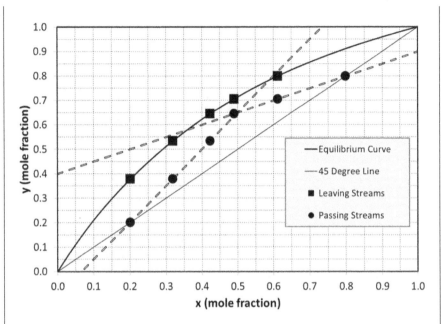

FIGURE 5.8

Benzene–toluene VLE at 1 atm. The top and bottom operating lines are drawn through the passing streams.

you will get the plot in Figure 5.8 (Excel "Multi-Stage Distillation/5-Stage Op Lines"). No matter the process specification, you will observe the lines go through the center of each point. The lines through the passing streams above the feed stage will be known as a top operating line (TOL) and through the passing streams below the feed stage will be known as a bottom operating line (BOL). Notice that they intersect at the feed composition, $z = 0.5$ mol B/mol.

Furthermore, you can connect the top stream with a horizontal line to the equilibrium stream (square), then a vertical line to the next passing stream (dot), and so on. Notice that it makes a perfect staircase. This is illustrated in Figure 5.9 (Excel "Multi-Stage Distillation/5-Stage Stairs"). Congratulations! You have just discovered the famous graphical design technique called the McCabe–Thiele method to be developed in Chapter 7. That, young engineer, is awesome!

e. Be awesome!

You can input all the equations in Excel "Multi-Stage Distillation/5-Stage DIY." Add F_i, z_i to the mole balances

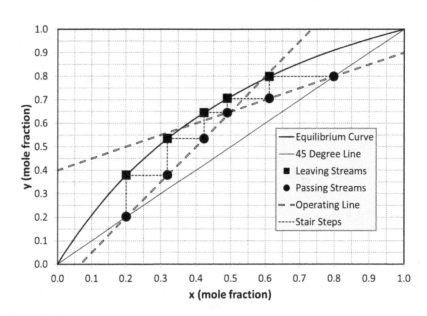

FIGURE 5.9
Benzene–toluene VLE at 1 atm. The stair step construction is demonstrated.

for Stage 2–4 to account for different feed locations. Just fill in F = 100 mol/s and z = 0.5 mol B/mol to either feed stage. If you select Stage 3 you should get the same results as above. Now, repeat the simulation for Stages 2 and 4 and compare results. It makes a difference and we'll find the reason soon!

5.3 Summary

Separation can be improved by using multiple equilibrium stages in series. Increasing the amount recycled back to the column as reflux also increases separation. The molar flow rates, compositions, and temperatures of all streams can be calculated using the same equations used for a single stage with the addition of a few process specifications. These simple equations are the total mole balance, component mole balance, and equilibrium relationships (Raoult's law for ideal solutions). A new process specification, EMO, required that total vapor and liquid molar flow rates were constant ($V_{i+1} = V_i$, $L_i = L_{i+1}$) unless material was added as feed or removed as product. The complexity of the analysis only increases because of the number – not difficulty – of equations to solve.

6

Mathematical Analysis of Distillation Columns

In the last chapter, we analyzed the performance of five equilibrium stages in series and observed a significant improvement in separation over single and two-stage processes. This is a type of analysis problem; we are given the process design (number of stages, position of feed, etc.) and operating conditions (pressure, reflux ratio, etc.) and asked how it will perform.

The temperature of each stage along with the flow rates and compositions of streams leaving and entering was determined. The governing equations were mole balances and equilibrium relationships. A degree of freedom (DOF) analysis revealed that no subsystem, individual stage, had enough information to solve the equations. We resorted to a "brute force" method by having the Excel Solver tool iterate for all remaining unknown variables at once. Even then, the problem was underspecified and the equimolal overflow (EMO) assumption was required as a process specification.

In this chapter, we will develop the more elegant stage-by-stage Lewis method (Lewis, 1922) to analyze a distillation column. As long as the EMO assumption can be applied, an energy balance is not necessary unless we need the heat duties of the condenser and reboiler.

6.1 Overall Material Balance

Figure 6.1 is a typical diagram of a distillation column and major associated equipment. In practice, this would be a single column with internal trays or sections of packing acting as stages. At the top is a total condenser (TC) that changes the vapor into a liquid, which is split into a product (distillate) and a recycle stream (reflux) that is returned to the top stage. At the bottom is a partial reboiler (PR) that evaporates part of the liquid to be returned to the bottom stage as a recycled vapor (boilup) with the remaining liquid taken as a product (bottoms).

If we draw a material balance envelope (MBE) around the entire process, the only streams that cross the boundary are the feed, distillate, and bottoms with their corresponding molar flow rates F, D, and B and mole fractions x_F, x_D, and x_B of the light component. If F, z, and B are specified, that leaves three unknowns

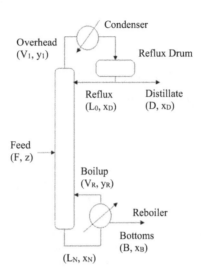

FIGURE 6.1
PFD of a typical distillation column.

(D, x_D, and x_B) and two equations (total and component mole balances). D is, of course, easily determined from the total mole balance, but there is no way of knowing *a priori* the product compositions. We can't use an equilibrium relationship because the distillate and bottoms are not in equilibrium – that only applies to streams leaving individual equilibrium stages. We could specify one of the compositions and solve for the other, but there is no guarantee that the process can achieve it. A more sophisticated analysis is required.

6.2 Lewis Method

Figure 6.2 is a simple process flow diagram (PFD) of a ten-stage distillation column with a TC and a PR. Most streams have not been labeled for simplicity; however, the vapor stream leaving any Stage i would be V_i with composition y_i and the liquid stream leaving any Stage i would be L_i with composition x_i. The liquid leaving the TC is split into a distillate product, D, and a reflux stream, L_0, according to a specified reflux ratio (L_0/D). Note that the composition of V_1, D, and L_0 must be the same since no separation is taking place in a TC or split point. The feed, F, to be separated is sent in this case to Stage 5. Some of the liquid leaving Stage 10 and fed to the PR is vaporized, V_R, (boilup) and the remaining liquid is the bottoms product, B. Notice that the products (D and B) are both taken as liquids, the easiest phase to physically handle. There are many possible configurations of distillation columns as will be discussed in Chapter 9.

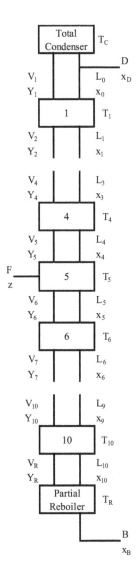

FIGURE 6.2
A simple PFD of a ten-stage distillation column with a TC and PR. Streams are labeled according to the stage they leave.

Calculation of flow rates, compositions, and temperatures (assuming a constant pressure) could be made in exactly the same way (brute force method) as was done for the five-stage column in Chapter 5. However, here we will develop a more elegant method. Certain specifications are made about the feed and products. Then the mole balance and equilibrium equations are solved stage-by-stage with the EMO assumption. This is called the Lewis method (Lewis, 1922).

EXAMPLE 6.1   Ten-Tray Distillation Column

A total of 100 mol/s of an equimolar, liquid mixture of benzene and toluene is fed to the fifth tray of a ten-tray distillation column equipped with a PR and a TC. The reflux ratio is 1.0 with equal amounts of distillate and bottoms produced. Assume EMO and perfectly efficient trays. We also assume the feed and reflux are saturated liquids (at their bubble points). If they were subcooled, they would condense some vapor, complicating the analysis.

a. Draw a labeled process flow diagram.
   As usual with the engineering approach to problem-solving, we'll break the problem down into simpler subsystems, starting with a MBE around the TC/split point as shown in Figure 6.3. Note that with F = 100 mol/s and an equal split between distillate and bottoms, D and B must each be 50 mol/s. Also, note that no separation takes place in the TC or split point so that $x_D = x_0 = y_1$.

b. Conduct a DOF analysis.
   The streams crossing the MBE are V_1, D, and L_0.

   Unknowns (4): V_1, x_D, L_0, T_C

   Equations (4):
      2 mole balances (total and B)

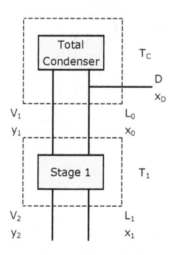

FIGURE 6.3
MBEs around the top portion of a distillation column. V, L, and D are vapor, liquid, and distillate molar flow rates (mol/s), respectively, and x and y are liquid and vapor mole fractions respectively.

1 equilibrium relationship (bubble point equation)

1 process specification (PS) (reflux ratio)

Since there are no vapor-phase compositions in equilibrium with liquid-phase compositions, we don't have the usual equilibrium relationships. However, Tc is related to x_D by the bubble point equation if the liquid is at it the bubble point. This looks properly specified with zero DOF – or is it? The total and benzene mole balances are not independent equations if all the mole fractions are equal. Prove this to yourself.

So there actually is one DOF and the unknowns cannot yet be determined. We can either move on to find a different subsystem with zero DOF (good luck with that!) or make another assumption. We will *guess* $x_D = 0.85$. Now there are three unknowns, three independent equations, and zero DOF. We can solve for the unknowns.

A subsequent MBE around Stage 1 involves streams V_1, L_1, V_2, and L_0. We already know V_1, y_1, L_0, and x_0, from the above calculation.

Unknowns (5): V_2, y_2, L_1, x_1, T_1

Equations (5):

2 mole balances (total and B)

2 equilibrium relationships (Raoult's law for B and T)

1 process specification (EMO)

There are zero DOF so all unknowns can be determined. Now we can proceed with stage-by-stage calculations down the column. Another way of constructing the MBE is to draw it around the entire top of the column so that the only streams crossing it are V_{i+1}, L_i, and D.

Is it really fair to guess x_D? Of course, I wouldn't think of cheating! At the very end of the calculation (at the PR) we'll check the overall benzene mole balance. Does $Fz = Dx_D + Bx_B$? If not, we'll choose another value for x_D and repeat the solution until the overall mole balance is satisfied. As a young student, I did this with pencil, paper, a slide rule, and no social life. You can use Excel "Lewis/10-Tray" and let the Solver tool do this iteration for you in seconds! Then you should go to the gym and get some exercise!

c. Write the equations and solve.

For the TC/split point MBE:

Guess: $x_D = 0.85 \, \text{mol B/mol}$

Total mole balance: $V_1 = D + L_0$

Bubble point equation: $x_D K_{CB} + (1-x_D)K_{CT} = 1$ (Solve for T_C by iteration)

PS reflux ratio: $R = L_0/D$

All calculations are done in Excel "Lewis/10-Tray." The results for the condenser/split point are shown in the table below. Guessed values are shaded.

Stage	L	x	V	y	T	K_B	K_T	BP
Condenser/split point	50	0.85	---	---	83.3	1.101	0.428	1.000
1			100	0.85				

Since streams are labeled by the stage they are leaving, V_1 and y_1 appear in the row for Stage 1.

For Stage 1, the equations are:

Total mole balance: $V_2 + L_0 = V_1 + L_1$

Benzene mole balance: $V_2 y_2 + L_0 x_0 = V_1 y_1 + L_1 x_1$

Benzene equilibrium: $y_1 = K_{1B} x_1$

Dew point equation: $y_1/K_{1B} + (1-y_1)/K_{1T} = 1$ (Use dew point because y_1 is known from Condenser/split point)

PS (EMO): $V_2 = V_1$

The new values from the Stage 1 calculations are shown below. The previous values from specifications and the condenser/split point are shaded. Note that the y_{i+1} values are always calculated from the Stage i benzene mole balance.

Stage	L	x	V	y	T	K_B	K_T	BP
Condenser/split point	50	0.85	---	---	83.3	1.101	0.428	1.000
1	50	0.69	100	0.85	87.0	1.230	0.485	1.000
2			100	0.77				

Table 6.1 shows the calculations after this first iteration. Note that the stage-by-stage calculation and the overall process mole balance don't agree. A new guess for x_D must be made. Try some of these calculations by hand.

TABLE 6.1

First Iteration of Lewis Method Design Calculations

Stage	L	x	V	y	T
TC	50	0.850			83.3
1	50	0.691	100	0.850	87.0
2	50	0.572	100	0.770	90.1
3	50	0.497	100	0.711	92.2
4	50	0.454	100	0.673	93.5
5	150	0.431	100	0.652	94.2
6	150	0.352	100	0.571	96.7
7	150	0.255	100	0.453	100.1
8	150	0.156	100	0.307	103.8
9	150	0.074	100	0.159	107.3
10	150	0.016	100	0.036	109.9
Reboiler	50	−0.021	100	−0.052	111.7

X_R by Overall Mol Bal = 0.150

$\Delta = -0.171$

TABLE 6.2

Final Iteration of Lewis Method Design Calculations

Stage	L	x	V	y	T
TC	50	0.891			82.4
1	50	0.761	100	0.891	85.3
2	50	0.652	100	0.826	88.0
3	50	0.574	100	0.771	90.0
4	50	0.523	100	0.732	91.5
5	150	0.492	100	0.707	92.4
6	150	0.464	100	0.683	93.2
7	150	0.420	100	0.642	94.5
8	150	0.356	100	0.576	96.6
9	150	0.274	100	0.479	99.4
10	150	0.187	100	0.357	102.6
Reboiler	50	0.109	100	0.226	105.7

X_R by Overall Mol Bal = 0.109

$\Delta = 0.000$

For the final solution, let x_D be a variable and the difference (Δ) between the stage-by-stage calculations and the overall process mole balance be a constraint. Table 6.2 shows the final values for temperature, flow rates, and compositions.

Note the improvement in separation ($x_B = 0.109$, $x_D = 0.891$) compared to a five-stage column with the same specifications ($x_5 = 0.201$, $y_1 = 0.799$) and a single stage ($x = 0.267$, $y = 0.465$).

Heat will be added to the PR and removed from the TC. We usually assume that the column (and its trays) is adiabatic. Adding the energy balance to each stage's calculations will allow us to drop the EMO assumption since we are not adding additional unknowns. We'll defer this calculation until Chapter 8.

d. Explore the problem.

Some finer points that only you would understand: The distillation column's TC does not count as a stage because no separation occurs with a product of only one phase. The PR actually does produce two phases and counts as a stage. Therefore, a ten-tray distillation column is actually an 11-stage column (count the number of steps touching the equilibrium curve). The five-stage column had a partial condenser at the top, a PR at the bottom, and three trays in the middle. Therefore, it actually had five equilibrium stages.

Let's try the same graphical analysis of these results that we made for the five-stage column. In Figure 6.4, the "leaving

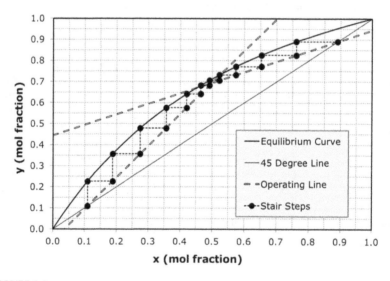

FIGURE 6.4

Vapor–liquid equilibrium for benzene and toluene at 1 atm. Vapor and liquid compositions for streams leaving or passing between each of the 11 stages in Example 6.1 are plotted. A straight operating line can be drawn through the passing stream compositions in the upper and lower stages. The leaving and passing stream compositions can be connected by horizontal and vertical lines resembling a stair case.

streams" (x_i, y_i) and the passing streams (x_i, y_{i+1}) are plotted. All leaving streams lie on the equilibrium line because they are in equilibrium (an assumption we can relax later to account for tray efficiency). The passing streams seem to lie on one straight line above the feed stage and another straight line below the feed stage. These are called the top and bottom operating lines (TOL and BOL). They intersect at the feed composition.

In addition, we can draw a series of horizontal and vertical lines between the passing and leaving streams that look very much like stair steps. The McCabe–Thiele method (McCabe and Thiele, 1925) for distillation column design is a graphical technique based on this observation. The method inspired the #1 chemical engineering themed song "Steppin' Off The Stages" by Professor Doobie 'Doghouse' Wilson (DoobieDoghouseWilson.com) (Lane, 2008).

e. Be awesome!

Use Excel "Lewis/10-Tray" to examine the effect of reflux ratio on product compositions. Try increasing the ratio to 2.0 and use Solver to update the flow rates and compositions. The xy graph will be updated also. This increases the internal liquid and vapor flow rates in the column. It also increases the heating and cooling requirements in the reboiler and condenser. As a design engineer, what do you expect will be the consequences for capital and operating costs?

TRY THIS AT HOME 6.1 Fun with a Ten-Tray Distillation Column

A total of 100 mol/s of a liquid mixture with 40 mol% benzene and 60 mol% toluene is fed to the fifth tray of a ten-tray distillation column equipped with a PR and a TC. The reflux ratio is 2.0 with twice as much distillate as bottoms produced (D = 66.7 mol/s). Assume EMO and perfectly efficient trays. We also assume the feed and reflux are saturated liquids (at their bubble points). Determine the composition of the bottoms and distillate.

You already have Excel "Lewis/10-Tray" set up, so simply input these new specifications, run Solver, and look at Excel "Lewis/Stair Steps." Notice that the bottoms are very pure (0.997 mol T/mol) but the distillate is not so pure (0.598 mol B/mol).

Change the feed stage and decrease the distillate flow rate to obtain at least a 90 mol% B distillate and a 10 mol% B bottoms. Run as many process specifications – including reflux ratio – as you can and try to explain the results. Note how the operating lines and stair steps change with each specification.

6.3 Summary

Distillation columns commonly have many stages. The product compositions, x_D and x_B, are not in equilibrium so they must be determined by a stage-by-stage analysis of the entire column. With typical specifications, no stage or subsystem has zero DOF. In the last chapter, we got around that problem with the five-stage column by using the "brute force" method, solving all equations at once. The Lewis method is a bit cleverer. We make a reasonable guess for the product composition and conduct the stage-by-stage analysis. At the end, we see if the compositions satisfy the overall mole balance within a certain tolerance. If not, we just go back and make a new guess. Excel's Solver tool makes this easy!

7

Graphical Design of Distillation Columns

In Chapters 5 and 6, we found that plotting and connecting leaving and passing stream compositions in a multi-stage distillation column formed a staircase on the xy graph. This is the basis of the McCabe–Thiele method, a well-known graphical technique to design a distillation process. Both the analytical Lewis method and the graphical McCabe–Thiele method require assuming equimolal overflow (EMO).

For this discussion, we'll assume a bubble point benzene–toluene feed with $F = 100.0\,\text{mol/s}$, $z = 0.500\,\text{mol B/mol}$, $D = 50.0\,\text{mol/s}$, $x_D = 0.891\,\text{mol B/mol}$, and $R = 1.00$. (Note that this happens to be the final solution for the ten-tray distillation column in Example 6.1.) This method will allow us to determine the number of stages and optimum feed stage, which should turn out to be exactly the same. I highly recommend you to print some blank xy diagrams from Excel "McCabe-Thiele/xy" and follow along with each step.

Here is a preview of the method. Start with the xy graph showing the equilibrium curve. Draw the top operating line (TOL) that represents mole balances across the top of the column and includes the coordinates of passing streams above the feed stage. Then draw the bottom operating line (BOL) that represents mole balances across the bottom of the column and also includes the coordinates of passing streams below the feed stage. The intersection of the TOL and BOL also includes a line representing the feed, as we will see. Finally, the stair steps are constructed between the equilibrium line and TOL starting at x_D. (You could also start at the bottom with x_B.) After crossing the TOL/BOL intersection, the steps continue using the BOL and that stage is the optimum feed location. The number of required stages is determined when the bottoms concentration reaches or passes the value determined by an overall mole balance. OK – now how does this work?

7.1 Top Operating Line

Figure 7.1 is a simplified process flow diagram (PFD) of the top of the distillation column. Let's start with the total condenser (TC). Take your pencil and draw a material balance envelope (MBE) around the TC and the distillate/reflux split point so that only streams V_1, D, and L_0 cross the boundary. On the xy graph in Figure 7.2, mark the composition of the passing

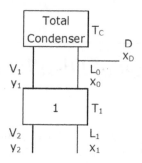

FIGURE 7.1

A PFD of the top portion of a distillation column. V, L, and D are vapor, liquid, and distillate molar flow rates (mol/s), respectively, and x and y are liquid and vapor mole fractions (mol B/mol), respectively.

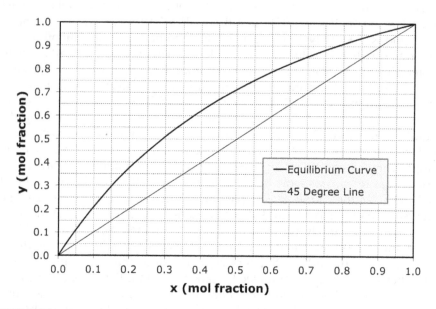

FIGURE 7.2

Vapor-liquid equilibrium (VLE) of benzene–toluene at 1 atm for use in developing the McCabe–Thiele design method based on the system specified in Example 6.1.

streams (L_0 and V_1) on the 45° line. That point is $(x_0, y_1) = (x_D, y_1) = (0.891, 0.891)$ since there is no change in composition in either the condenser or split point.

The benzene mole balance equation is

$$V_1 y_1 = L_0 x_0 + D x_D \qquad (7.1)$$

and can be put into a straight-line form

$$y_1 = \left(L_0/V_1\right) x_0 + \left(D/V_1\right) x_D \qquad (7.2)$$

This equation is a relationship between the streams passing between the TC/split point and Stage 1. Now draw an MBE around the TC and Stage 1 in Figure 7.1. Only streams V_2, D, and L_1 cross this boundary. The benzene mole balance in straight-line form is

$$y_2 = (L_1/V_2)x_1 + (D/V_2)x_D \qquad (7.3)$$

The EMO assumption is that $L_0 = L_1 = L_2 = \ldots$ and $V_1 = V_2 = V_3 = \ldots$ unless material is added or removed, which will happen someplace down the column at the feed stage. So above the feed stage let's use the symbols L and V for the constant liquid and vapor molar flow rates. For any pair of passing streams, the compositions (x_{i-1}, y_i) will be on the same straight line with a constant slope, L/V, and intercept, $(D/V)x_D$. This is called the TOL and is the benzene mole balance around the top of the column down to Stage i. Only streams D, V_i, and L_{i-1} cross this MBE.

$$y = (L/V)x + (D/V)x_D \qquad (7.4)$$

x and y are understood to be the compositions of liquid and vapor passing streams.

At $x_0 = x_D$ (the passing streams, V_1 and L_0, below the TC)

$$y_1 = (L/V)x_D + (D/V)x_D = [(L+D)/V]x_D = (V/V)x_D = x_D$$

We used the fact that $L_0 + D = V_1$. This exercise only showed that our original mark on the xy graph does indeed fall on the TOL. With one point on the line located (x_D, y_1), we can complete the line with either the slope or the intercept. For these specific operating conditions:

$$L = L_0 = R \times D = 1 \times 50 = 50$$

$$V = V_1 = L_0 + D = 50 + 50 = 100$$

$$\text{Slope} = (L/V) = 50/100 = 0.5$$

$$\text{Intercept} = (D/V)x_D = (50/100)0.891 = 0.446$$

In Figure 7.2, place a mark on the points (0.691, 0.791), (0.491, 0.691), and (0.291, 0.591), moving down 0.1 units and left 0.2 units from the original mark (slope of 1/2), and draw a line. If you were very careful, it will intercept the vertical axis at ~0.446. Alternatively, you could have simply drawn a line from (0.891, 0.891) to the intercept (0, 0.446). The composition of all passing streams above the feed stage will be on this line.

Another form of the TOL is found from the definition of reflux ratio, $R \equiv L_0/D$ and a little bit of algebra.

$$y = \frac{R}{1+R}x + \frac{1}{1+R}x_D \tag{7.5}$$

I challenge you to derive this! (You might try "reverse engineering" the equation by substituting L_0/D for R and rearranging to Eq. 7.4.) Of course, the slope and intercept will be the same.

7.2 Bottom Operating Line

We start with the bottom of the column shown in Figure 7.3. Take your pencil and draw an MBE around the partial reboiler (PR) so that the streams V_R, L_{10}, and B cross the boundary. Streams V_R and B are leaving streams whose compositions must lie on the equilibrium line at (x_B, y_R). x_B is easily determined from an overall process benzene mole balance:

$$Fz = Dx_D + Bx_B \tag{7.6}$$

So $x_B = (Fz - Dx_D)/B = (100 \times 0.5 - 50 \times 0.891)/50 = 0.109$

Mark the equilibrium line in Figure 7.2 at $x_B = 0.11$. $y_R = 0.23$ at this point since these are both leaving streams (don't expect three place precision on these graphs!).

We can develop an equation for the BOL in a similar manner as the TOL. A benzene mole balance around the PR is

$$L_{10}x_{10} = V_R y_R + Bx_B \tag{7.7}$$

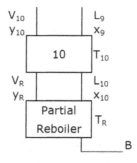

FIGURE 7.3
A PFD of the bottom portion of a distillation column. V, L, and B are vapor, liquid, and bottoms molar flow rates (mol/s), respectively, and x and y are liquid and vapor mole fractions, respectively.

or in straight-line form is

$$y_R = (L_{10}/V_R)x_{10} - (B/V_R)x_B \qquad (7.8)$$

This equation is a relationship between the streams passing between Stage 10 and the PR. On Figure 7.3, draw another MBE around both the PR and Stage 10 so that only streams V_{10}, L_9, and B cross the boundary. The benzene mole balance in a straight-line form is

$$y_{10} = (L_9/V_{10})x_9 - (B/V_{10})x_B \qquad (7.9)$$

Again, the EMO assumption is that $L_{10} = L_9 = L_8 = \ldots$ and $V_R = V_{10} = V_9 = \ldots$ unless material is added or removed, which will happen someplace up the column at the feed stage. So below the feed stage let's use the symbols L' and V' for the constant liquid and vapor molar flow rates. For any pair of passing streams below the feed stage, the compositions (x_{i-1}, y_i) will be on the same straight line with a constant slope, L'/V', and intercept, $-(B/V')x_B$. This is called the BOL and is a benzene mole balance around the bottom of the column up to Stage i. Only streams V_i, L_{i-1}, and B cross this boundary.

$$y = (L'/V')x - (B/V')x_B \qquad (7.10)$$

x and y are understood to be the compositions of passing liquid and vapor streams below the feed stage.

If we assume a bubble point liquid feed at the optimal feed stage (more about this later), it will only add to the liquid stream so that with these specific operating conditions

$$L' = L + F = 50 + 100 = 150$$

$$V' = V = 100$$

The slope and intercept of the BOL are

$$\text{Slope} = L'/V' = 150/100 = 3/2$$

$$\text{Intercept} = -(B/V')x_B = -(50/100) \times 0.109 = -0.055$$

We need a point on the BOL, but there is no passing stream associated with the bottoms stream, B. Also, the intercept with the vertical axis is unfortunately off the chart (negative). But, we can just pick any x value and calculate a y value using the equation. Might as well pick x_B:

$$y = (L'/V')x_B - (B/V')x_B = ((L' - B)/V')x_B = (V'/V')x_B = x_B$$

So the bottoms composition on the 45° line is on the BOL. On Figure 7.2, mark that point (0.109, 0.109) and then go up 0.3 units and right 0.2 units (slope of 3/2), mark the point (0.309, 0.409), and draw a line between the two points. This is the BOL.

Notice that the BOL intersects the TOL precisely at x = 0.5, the feed composition. This happens with a saturated liquid feed that completely adds to the liquid stream (F + L = L′). It is a consequence of the feed stage material balance represented by what is known as the feed or "q" line.

7.3 Feed Stage "q" Line

The TOL and BOL obviously intersect at some point. This is where (x, y) is the same for both equations:

$$TOL: yV = Lx + Dx_D$$

$$BOL: yV' = L'x - Bx_B$$

Subtracting the BOL from the TOL yields the feed line, which is a straight line including the intersection point.

$$y(V - V') = x(L - L') + (Dx_D + Bx_B) = x(L - L') + Fz \tag{7.11}$$

In the last part of Eq. 7.11, we used the overall benzene mole balance $Dx_D + Bx_B = Fz$. Note that $V - V' = V_F$ and $L' - L = L_F$ where V_F and L_F are the flow rates of vapor and liquid, respectively, in the feed. If it is a bubble point liquid, $L_F = F$ and $V_F = 0$. If it is a dew point vapor, $V_F = F$ and $L_F = 0$. If the feed is a mixed vapor/liquid, the feed stream contributes to both V and L′. (We will consider subcooled feeds in Chapter 9 because energy balances must be used.) Also, this is a bit of a simplification because it depends on the thermodynamic state of the other streams entering and leaving the stage. The feed line becomes

$$y = -(L_F/V_F)x + (F/V_F)z \tag{7.12}$$

This is a vertical line if the feed is a saturated liquid ($V_F = 0$) and a horizontal line if the feed is a saturated vapor ($L_F = 0$). The parameter "q" represents the quality of the feed stream in the sense of the fraction that is liquid. We write $q = (L' - L)/F$ or L_F/F so that if the feed is a bubble point liquid, $q = 1$, if the feed is a dew point vapor, $q = 0$, and if it is a mixed phase feed, $0 < q < 1$. It can be worked into the feed line equation with a little bit of algebra. See if you can derive this.

$$y = -(q/(1-q))x + (1/(1-q))z \tag{7.13}$$

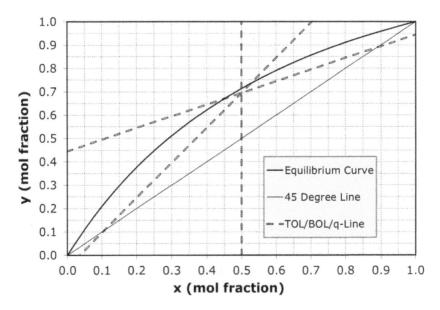

FIGURE 7.4
VLE of benzene–toluene at 1 atm with the TOL, BOL, and feed or q-line for the system specified in Example 6.1. Make sure your lines on Figure 7.2 look like this.

This is another form of the feed line and commonly called the q-line. It contains the point (z, z) (prove this!) and has a vertical or infinite slope for $q = 1$ (saturated liquid feed) and a horizontal or zero slope for $q = 0$ (saturated vapor feed) as we saw above. Now we have another way to draw the BOL. Draw the q-line and then connect the intersection of the q-line with the TOL to the bottoms composition (x_B, x_B).

Figure 7.4 is the xy graph with the TOL, BOL, and q-line. Make sure the graph you have been constructing in Figure 7.2 looks just like this so far.

7.4 Steppin' Off the Stages

Now comes the really fun part – steppin' off the stages! y_1 ($=x_D$) is in equilibrium with x_1 so draw a horizontal line on Figure 7.4 from $(x_D, y_1) = (0.891, 0.891)$ to $(x_1, y_1) = (0.762, 0.891)$ on the equilibrium curve. x_1 is related to y_2 by a material balance (the operating line), so draw a vertical line from $(x_1, y_1) = (0.762, 0.891)$ to $(x_1, y_2) = (0.762, 0.827)$ on the TOL. One more time? y_2 is in equilibrium with x_2 so draw a horizontal line from $(x_1, y_2) = (0.762, 0.827)$ to $(x_2, y_2) = (0.653, 0.827)$ on the equilibrium curve. x_2 is related to y_3 by a material balance, so draw a vertical line from $(x_2, y_2) = (0.653, 0.827)$ to

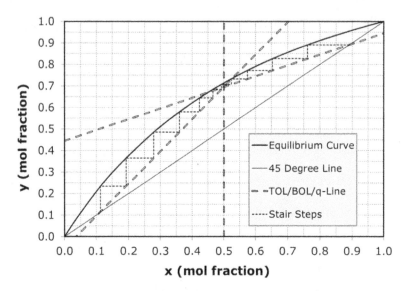

FIGURE 7.5
Completed McCabe–Thiele design of the system specified in Example 6.1.

$(x_2, y_3) = (0.653, 0.772)$ on the TOL. Think of each point on the equilibrium line as representing an equilibrium stage. (x_1, y_1) is Stage 1, (x_2, y_2) is Stage 2, etc.

Please continue – using a straight edge and being very careful – on down the column. When you get to the TOL/BOL/q-line intersection, you can travel further with each step by going down to the BOL rather than the TOL. That will be the optimal feed stage (should be Stage 5 in this example). Continue past the feed stage to the BOL and you should reach (x_B, x_B) at Stage 11 (the PR). If you've been a very careful and talented chemical engineer, the graph should look exactly like Figure 7.5. If your steps didn't land exactly on 0.11 don't worry; the steps are rather small near the feed stage with this low-reflux ratio. Any imprecision in the stair stepping is magnified.

We used the exact results from Example 6.1 for specifications and landed on the bottoms concentration in 11 stages (including the PR) with feed on the fifth stage. Normally, it will not work this nicely because we will specify the performance and ask "how many stages" and "where should we put the feed"? This is a design problem as illustrated in the next couple of examples.

EXAMPLE 7.1   McCabe–Thiele Design
with a Saturated Liquid Feed

A total of 100 mol/s of a saturated liquid mixture of 45 mol% benzene and 55 mol% toluene is fed to the optimal tray of a distillation column operating at 1 atm and equipped with a TC and PR. About 40 mol/s of a saturated liquid distillate with 0.95 mole B/mole is produced using a reflux ratio R = 2.0. Assume EMO and 100% efficient trays. How many trays are required and what is the optimal feed tray? Print out an xy graph and follow along.

a. Draw a labeled process flow diagram and
b. Conduct a degree of freedom analysis.

Figure 7.6 is the best we can do since we don't know the number of stages or where the feed should be. However, it does help visualize the problem. It shows N trays and feeding to Tray i. We have already seen that enough information was given to design this process.

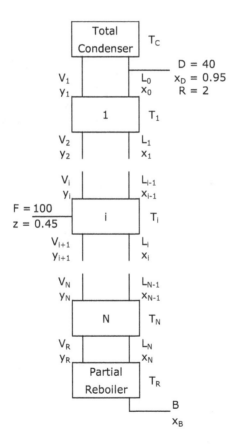

FIGURE 7.6
PFD of a distillation column with an unknown number of trays and feed position. This is a typical design situation.

c. Write the equations and solve.

First, we will draw the TOL on an xy graph. (Did you print one?) Mark the distillate composition on the 45° line (0.95, 0.95). Next mark the y-intercept = $x_D/(1 + R) = 0.95/3 = 0.32$. Take a long straight edge and connect the points.

Next, draw the q-line. It will be a vertical line (saturated liquid feed) starting at the feed composition on the 45° line (0.45, 0.45).

Next, draw the BOL. Mark the bottoms composition on the 45° line (0.12, 0.12) (this is calculated using the overall mole balance – try it!) and connect that point with the TOL/q-line intersection.

You can also draw the TOL and BOL lines with one point and a slope if you prefer. Make sure your xy graph looks like Figure 7.7 before continuing.

Finally, carefully step off the stages with a series of horizontal moves to the equilibrium line and vertical moves to the operating line (see Figure 7.8). At Stage 6, we cross the q-line and get better separation by dropping down to the BOL. So this will be our feed stage. Continuing with the BOL, we find that Stage 10 exceeds the desired bottoms composition of 0.12 mol B/mol. So we need ten stages, or nine equilibrium trays and a PR. Remember that the TC provides no separation and is not an equilibrium stage.

It would be pure coincidence if we "landed" exactly on $x_B = 0.12$. Some overshoot is a good thing; it builds in some room

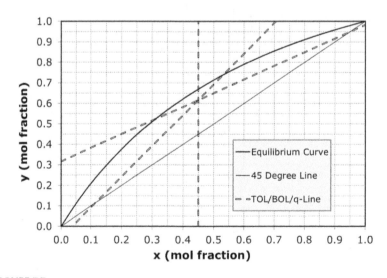

FIGURE 7.7
VLE for toluene–benzene at 1 atm showing the TOL, BOL, and feed or q-line for Example 7.1.

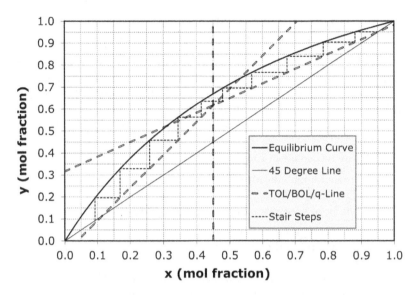

FIGURE 7.8
VLE for toluene–benzene at 1 atm showing the completed McCabe–Thiele design for Example 7.1. This separation requires ten stages with Tray 6 as the optimal feed stage. Use nine equilibrium trays and a PR.

for error. We could always reduce the reflux ratio slightly to produce a bottoms closer to the target. (You can easily do this in Solver by adding $x_B = 0.117$ as a constraint and adding R as a variable. In this case R = 1.908.)

The required heat added at the reboiler and removed at the condenser can be calculated with enthalpy balances. Not only that, enthalpy balances at each stage will allow us to drop the EMO assumption. We will do this in Chapter 8.

d. Explore the problem.

Back off the reflux ratio to 1.91 (as suggested above) and repeat the McCabe–Thiele analysis graphically and using Excel "McCabe-Thiele/Liquid Feed." This should come much closer to $x_B = 0.12$ with ten equilibrium stages and feeding to Stage 6. Alternatively, you can increase the reflux ratio and hit the target x_B with nine equilibrium stages.

e. Be awesome!

Copy the spreadsheet for a column with ten equilibrium trays and a PR in Excel "Lewis/10-Tray" and modify it to have one less equilibrium tray. Verify that R = 1.91 produces $x_B = 0.12$ with the above specifications.

TRY THIS AT HOME 7.1   McCabe–Thiele Design with Saturated Vapor Feed

A total of 100 mol/s of a saturated (dew point) vapor consisting of 60 mol% benzene and 40 mol% toluene is fed to a distillation column containing equilibrium trays, a TC, and a PR. A reflux ratio of 2.0 produces a 65 mol/s of a distillate with 0.89 mol B/mol. How many trays are required and which one will be the optimum feed tray?

  a. Draw a labeled PFD and
  b. Conduct a DOF analysis.

   The PFD (Figure 7.6) and DOF in Example 7.1 are the same here except for the specified values.

  c. Write the equations and solve.

   Print a copy of the xy diagram from Excel "McCabe-Thiele/xy."
   Draw the TOL.
   Draw the q-line for this saturated vapor feed. This will be a
    horizontal line that goes through (z, z).
   Draw the BOL.
   Step off the stages.

   After you have done this yourself, compare your result to the McCabe–Thiele diagram in Figure 7.9. The TOL

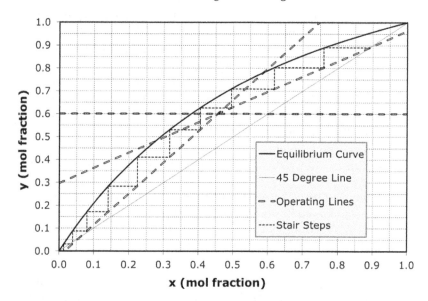

FIGURE 7.9
VLE of benzene and toluene at 1 atm. This is the completed McCabe–Thiele diagram for Try This at Home 7.1.

> y-intercept is 0.297. The process requires nine equilibrium
> stages (eight trays and the PR) with feed to Stage 4.
>   How do you suspect the overall energy balance of this
> system is affected by feeding a vapor rather than a liquid?
>
> d. Explore the problem.
>   Suppose the feed is actually an equimolar mixture of
> vapor and liquid. Now the q-line has a slope of –1. Repeat
> the analysis and explain your observations.
>
> e. Be awesome!
>   Copy and modify Excel "McCabe-Thiele/Liquid Feed" to
> account for a mixed vapor/liquid feed.

7.5 Limiting Conditions

There is a limit on what a distillation column can do even with an infinite
number of stages or an extremely high reflux ratio. These limiting conditions bracket what we can expect in our design. These conditions are called
total reflux and minimum reflux. Suppose the desired distillate is $x_D = 0.9$
and the desired bottoms is $x_B = 0.1$. The feed will be a saturated liquid with
composition $z = 0.5$.

7.5.1 Total Reflux

At this condition, the feed is turned off, all distillate is returned as reflux,
and all bottoms are returned as boilup. The system is at steady state with
no material entering or leaving the overall process. In this case, $L_0 = V_1$ so
the slope of both the TOL and BOL is $L_0/V_1 = V_R/L_N = 1$ and follows the 45°
line. This is shown in Figure 7.10. Starting at $(x_D, x_D) = (0.9, 0.9)$, construct
a McCabe–Thiele diagram by stepping off stages between the equilibrium
curve and 45° line. Go ahead and try this. You should find five stages, or four
equilibrium trays and a PR. Go ahead and draw this on Figure 7.10 or print
a copy of the xy diagram from Excel "McCabe-Thiele/xy" and give it a try.

7.5.2 Minimum Reflux

Now we will decrease the reflux causing the slope L/V to decrease. The
BOL is drawn from $(x_B, x_B) = (0.1, 0.1)$ to the intersection of the TOL and feed
line. Eventually, that intersection will touch the equilibrium curve as seen
in Figure 7.10. That is a pinch point and the number of stages becomes infinite. For this benzene–toluene system that point on the equilibrium curve

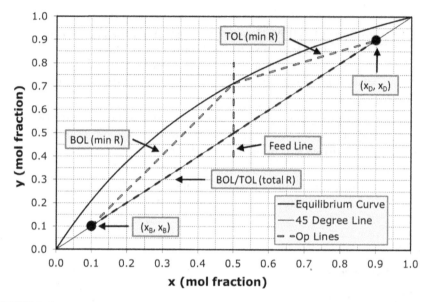

FIGURE 7.10
Demonstration of total and minimum reflux.

at $z = 0.5$ is $(0.5, 0.714)$. It is pointless to attempt constructing the stair steps. However, we can learn that the slope L/V is

$$\frac{L}{V} = \frac{\text{rise}}{\text{run}} = \frac{(0.9 - 0.714)}{(0.9 - 0.5)} = 0.465$$

Using Eq. 7.5 and a touch of algebra, this is a minimum reflux ratio of

$$R_m = \frac{0.465}{1 - 0.465} = 0.92$$

Normally, a percentage of R_m is specified, for instance "set the reflux ratio as 150% of the minimum."

7.6 A Moment of Reflection

You are now well equipped to analyze (using the mathematical Lewis method) or design (using the graphical McCabe–Thiele method) a standard distillation process including equilibrium stages, a TC, and a PR. You feel ready to go out and conquer the world of chemical separations – or at least ace the next exam. Patience, young chemical engineering padawan! There is so much more to learn.

1. What if we need to determine the heat removed in the condenser or heat added in the reboiler?
2. What if EMO is not a valid assumption?
3. What if the feed is a superheated vapor or (more likely) subcooled liquid?
4. What if the separation in the trays does not reach equilibrium?
5. What if the internals consist of packing rather than discrete trays?
6. What if the reflux is subcooled?
7. What if there are more than two chemicals in the feed?
8. What if the mixture is not ideal, i.e., equilibrium cannot be described by Raoult's law?
9. How tall and wide does the column need to be?
10. What affects the operability of the column?

We'll learn how to handle all those questions in upcoming chapters. There are even more questions that we'll leave to your instructor's supplementary material, an advanced course, vendors, and other disciplines. For instance:

1. What kind of trays or packing should be used?
2. What should the wall thickness be?
3. What is the pressure drop in the column?
4. How should the condenser and reboiler be designed?
5. What material of construction should be used?
6. What kind of control system should be used?
7. How is non-steady-state behavior such as startup, shutdown, and process upsets analyzed?
8. How are hazards identified and controlled in the process?
9. How is the physical layout of the process designed?
10. What are the capital and operating costs?

Can you think of other considerations? There are still many more. See if you can think of five more major questions and compare with your study group.

1.
2.
3.
4.
5.

Now you start to understand why chemical engineers are in high demand and get paid the big bucks! Because they are awesome!

7.7 Summary

In previous chapters, we "discovered" the McCabe–Thiele graphical method for distillation column design. The leaving stream compositions from each stage lie on the equilibrium line. The passing stream compositions between stages lie on an operating line. We used mole balances to derive those operating lines. The streams can be identified by a series of stair steps between the equilibrium curve and operating lines. The feed tray should be located where the largest change in concentration can occur by switching from the TOL to the BOL. The number of stages required is the number of steps required to achieve the specifications.

8

Energy Balances for Distillation Columns

Both the mathematical Lewis (Chapter 6) and graphical McCabe–Thiele (Chapter 7) methods for distillation column analysis and design required the assumption of equimolal overflow (EMO). This means that the vapor and liquid flow rates are constant except where material is added or removed; that is, typically at the feed stage, reboiler, and condenser. By adding energy balance equations for each stage, the EMO process specification is no longer necessary. We'll call this the "Rigorous" method.

8.1 Individual Stage Material and Energy Balances

Before looking at a specific example, it will be instructive to look at the equations governing an individual stage. The feed stage is shown in Figure 8.1, but for any other stage, simply ignore the feed stream.

The process variables are the flow rates, compositions, temperatures, and the amount of heat transferred. Pressure will be specified and assumed constant, leaving 15 variables. The equations relating these variables are two mole balances, two equilibrium relationships, an energy balance, and any

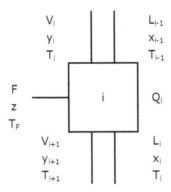

FIGURE 8.1
Labeled process flow diagram of the feed stage. F, V, and L are the stream molar flow rates (mol/s), x, y, and z are the mole fractions (mol B/mol), T is the temperature (°C), and Q is the heat transferred (kJ/s). Stream variables are subscripted according to the stage they leave. For other stages, simply leave off the feed stream.

appropriate process specifications. Recalling the stage-by-stage analysis using the Lewis method, V_i, y_i, L_{i-1}, x_{i-1}, and T_{i-1} are known from the analysis of the stage above, Stage $i-1$. Although not required before, we will have calculated T_i because the energy balance for Stage $i-1$ required that the temperature of stream V_i be known. Also, we usually completely specify the feed stream flow rate, temperature, and composition. That leaves six unknowns: V_{i+1}, y_{i+1}, T_{i+1}, L_i, x_i, and Q. If the column is perfectly insulated, we have an adiabatic stage, and simply define $Q \equiv 0$, resulting in five equations, five unknowns, and zero degrees of freedom (DOF).

8.2 Rigorous Analysis

The analysis should be almost the same as the Lewis method except adiabatic stages are assumed rather than EMO. Guess a value for x_D and then start at the top with the total condenser (TC)/split point and work on down the column. At the bottom, check to see if the overall mole balance is satisfied. If not, make a new guess for x_D, rinse, and repeat!

A word about the "old days." Before the rise of computers, a graphical technique that accounted for energy balances (and did not use the EMO assumption) was developed called the Ponchon–Savarit method. It is a bit complicated and few separations professors could even agree on the pronunciation, much less teach it. Now you will see that the rigorous method really isn't all that difficult with the aid of a spreadsheet.

**EXAMPLE 8.1   Distillation Column
Analysis Using Energy Balances**

A total of 100 mol/s of an equimolar benzene–toluene mixture (92°C, 1 atm, liquid) is fed to the fifth tray of a ten-tray distillation column equipped with a TC and partial reboiler (PR). Equimolar flow rates of distillate and bottom products are produced. The reflux ratio, R, is 2.0. Assume that all trays are adiabatic and 100% efficient [the leaving streams are truly at vapor-liquid equilibrium (VLE)]. Determine all unknown temperatures, flow rates, and compositions. Note that these are the specifications used in Examples 6.1 and 7.1 so that we can compare results.

a. Draw a labeled process flow diagram.
    The column, TC, and PR are all featured on the process flow diagram (PFD) shown in Figure 8.2.

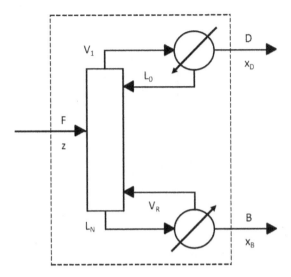

FIGURE 8.2
An MBE around the overall distillation process.

b. Conduct a degree of freedom analysis (overall process).

An overall material balance envelope (MBE) is drawn around the whole process so that only streams F, D, and B cross the line. We could add an energy balance to this analysis if we wish to calculate the heat removed at the TC (Q_C) and the heat added to the PR (Q_R), but for now we'll keep this simple.

Unknowns (4): D, x_D, B, and x_B

Equations (3):

2 mole balances (total and B)

1 process specification (equal product molar flow rates)

This leaves one DOF. Let's invoke the process specification right now to find that $D = B = 50$ mol/s. Please don't tell me you used a calculator for that! x_D and x_B cannot be related by equilibrium and arbitrarily specifying one will not necessarily be achievable in this ten-tray column with a reflux ratio of 2. (Note that using the McCabe–Thiele method, we did specify x_D, but in that case we didn't have a fixed number of stages, but rather the number of required stages was the question.) So a stage-by-stage method like the Lewis method must be employed. Let's try again!

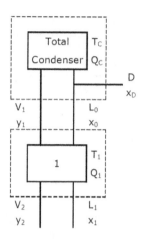

FIGURE 8.3
Process flow diagram of the top of the column in Example 8.1 including the TC and first tray.

a. Draw a labeled process flow diagram.
    As usual with the engineering approach to problem-solving, we'll break the problem down into simpler subsystems, starting with an MBE around the TC/split point as shown in Figure 8.3.

b. Conduct a degree of freedom analysis.

TC/Split Point
The streams crossing the TC/split point MBE are V_1, D, and L_0. We already calculated D, but we don't know V_1, L_0, T_C or the compositions (all the same). We will also need T_1 in order to calculate the enthalpy of stream V_1 for the energy balance.

Unknowns (6): V_1, T_1, x_D, L_0, T_C, Q_C

Equations (5):

   1 mole balances (total, the individual mole balances are not independent)

   1 energy balance

   2 equilibrium relationships (bubble point (BP) equation for T_C and dew point (DP) equation for T_1)

   1 process specification (reflux ratio)

Just like the Lewis method (Example 6.1), there is one DOF and the unknowns cannot yet be determined. We can make one more assumption, guessing $x_D = 0.85$. Now there are five unknowns, 5 independent equations, and zero DOF. We can solve for the unknowns.

Stage 1

A subsequent MBE around Stage 1 involves streams V_1, L_1, V_2, and L_0. We already know T_1, V_1, y_1, T_C, L_0, and x_0, from the above calculation. Again, we will need T_2 to calculate the enthalpy for stream V_2 for the energy balance.

Unknowns (6): V_2, T_2, y_2, L_1, x_1, Q_1

Equations (6):

> 2 mole balances (total and B)
>
> 1 energy balance
>
> 2 equilibrium relationships (B phase equilibrium, DP equation to calculate T_2)
>
> 1 process specification (adiabatic, $Q_1 = 0$)

There are zero DOF so all unknowns can be determined. Now we can proceed with stage-by-stage calculations down the column. As with the Lewis method, if the overall benzene mole balance is not satisfied at the end, we guess another value for x_D and repeat these calculations until it is right. At times like these you really start to appreciate your growing expertise in Excel and the use of Solver to do these iterations!

c. Write the equations and solve.

TC/Split Point

Guess: $x_D = 0.85$ mol B/mol

Total mole balance: $V_1 = D + L_0$

BP equation: $x_D K_{CB} + (1 - x_D)K_{CT} = 1$ (solve for T_C by iteration)

DP equation: $y_1/K_{1B} + (1 - y_1)/K_{1T} = 1$ (remember that $y_1 = x_D$. Solve for T_1 by iteration)

Process specification: $R = L_0/D$

Energy balance: $Q_C = H_D + H_{L_0} - H_{V_1}$

The enthalpy of a stream, H, is the product of the specific enthalpy of each species at the stream temperature and phase, the molar flow rate, and the composition. The distillate enthalpy, for example, is

$$H_D = D\left(x_D \hat{H}_{BD} + (1 - x_D)\hat{H}_{TD}\right) \qquad (8.1)$$

I chose the reference point for the enthalpy calculations as liquid at the normal boiling point of each substance although you can choose any reference condition that you fancy. The enthalpy of benzene in a liquid stream is simply

$$\hat{H}_{B,L} = \int_{T_B}^{T_S} C_{P,L} dT \tag{8.2}$$

T_B is the normal boiling point and T_S is the stream or stage temperature. The hypothetical path is B (liquid, T_B) → B (liquid, T_S) at constant pressure.

The enthalpy of benzene in the vapor phase is

$$\hat{H}_{B,V} = \Delta\hat{H}_V + \int_{T_B}^{T_S} C_{P,V} dT \tag{8.3}$$

The hypothetical path is B (liquid, T_B) → B (vapor, T_B) → B (vapor, T_S), all at constant pressure. The same calculation will be done for toluene with different physical properties, of course. Choosing this reference condition makes calculations somewhat simpler, but all streams will now have non-zero enthalpies.

All calculations are done in Excel "Rigorous/10-Tray." The row for the condenser/split point is shown in the table below. Since streams are labeled by the stage they are leaving, V_1 and y_1 appear in the row for Stage 1. L and V are flow rates (mol/s), x and y are mole fractions (mol B/mol), T is the temperature (°C), $\hat{H}_{iL}$ and $\hat{H}_{iV}$ are the specific enthalpies of each species in the liquid and vapor (kJ/mol), H_F, H_L, H_V are the stream enthalpies (kJ/s), and Q is the required heat transfer (kJ/s).

Stage	L	x	V	y	T	K_B	K_T	BP/DP
Condenser/ split point	100.0	0.850	---	---	83.3	1.101	0.428	1.000
1			150.0	0.850	87.0	1.230	0.485	1.000

Stage	$\hat{H}_{BL}$	$\hat{H}_{TL}$	$\hat{H}_{BV}$	$\hat{H}_{TV}$	H_F	H_L	H_V	Q
Condenser/ split point	0.46	−5.79	---	---	---	−47.7	---	−4766
1			31.45	30.42			4,695	

The guessed value for distillate composition (x_0 and y_1) is shaded. The temperatures, T_C and T_1, are found by iteration using the BP and DP equations, respectively. All the other values are calculated directly.

The heat removed from the TC is *roughly* the flow rate of vapor through the condenser multiplied by the opposite of the heat of vaporization of the mixture.

$$Q_C = -V_1 \Delta \hat{H}_{V,mix} = -V_1 \left(x_D \Delta \hat{H}_{V,B} + (1 - x_D) \Delta \hat{H}_{V,T} \right) = -4,676 \text{ kJ/s}$$

This number is approximate because the heat of vaporization is only exact at the normal boiling point for each species – not the condenser temperature. The value in the spreadsheet is based on a rigorous hypothetical path calculation. However, it is a worthwhile exercise to estimate values as a check on the calculations and they are close. Anyway, all these numbers are wrong at this point because we simply guessed x_D. No worries – we'll take care of that later.

Stage 1

Total mole balance: $V_2 + L_0 = V_1 + L_1$

Benzene mole balance: $V_2 y_2 + L_0 x_0 = V_1 y_1 + L_1 x_1$

Benzene equilibrium: $y_1 = K_{1B} x_1$

DP equation: $y_2/K_{2B} + (1 - y_2)/K_{2T} = 1$ (Use DP equation to find T_2)

Energy balance: $Q_1 = H_{V_1} + H_{L_1} - H_{V_2} - H_{L_0}$

Process specification (adiabatic): $Q_1 = 0$

The new values from the Stage 1 calculations are shown below.

Stage	L	x	V	y	T	K_B	K_T	BP/DP
Cond/SP	100.0	0.850	---	---	83.3	1.101	0.428	1.000
1	97.3	0.691	150	0.850	87.0	1.230	0.485	1.000
2			156.0	0.745	91.0	1.383	0.553	1.000

Stage	$\hat{H}_{BL}$	$\hat{H}_{TL}$	$\hat{H}_{BV}$	$\hat{H}_{TV}$	H_F	H_L	H_V	Q
Cond/SP	0.46	-5.79	---	---	---	-47.7	---	-4,766
1	1.01	-5.02	31.45	30.42	---	-83.5	4,695	0.000
2			31.86	30.93			4,659	

The previous values from the condenser/split point are shaded. The y_i values are always calculated from the Stage $i - 1$ benzene mole balance. The y_{i+1} values are calculated along with T_{i+1} at Stage i so that the enthalpy of stream V_{i+1} can be calculated for the energy balance.

Feed Stage

At the feed stage, the total mole balance, benzene mole balance, and energy balance equations will include the feed stream. With the analytical Lewis and graphical McCabe–Thiele methods, we assumed that the liquid and vapor portions of the feed simply added to the liquid and vapor flows, respectively. By calculating the feed stream enthalpy, we can account for its contributions to the liquid and vapor flow rates accurately.

The first step is to determine the BP and DP temperatures for the feed composition. If the feed is subcooled (below the BP temperature) then it is all liquid. In the rare case that the feed is superheated (above the DP temperature) then it is all vapor. If the temperature is in the two-phase region, the feed's vapor (V_F) and liquid (L_F) flow rates and compositions (y_F and x_F) can be determined with a flash calculation, just like you used in Chapter 3 for a single equilibrium stage.

The second step is to calculate the liquid and vapor phase-specific enthalpies of benzene ($\hat{H}_{BL}$ and $\hat{H}_{BV}$) and toluene ($\hat{H}_{TL}$ and $\hat{H}_{TV}$) at the feed temperature and pressure. The third step is to calculate the feed enthalpy.

$$H_F = V_F\left(y_B\hat{H}_{BV} + (1-y_B)\hat{H}_{TV}\right) + L_F\left(x_B\hat{H}_{BL} + (1-x_B)\hat{H}_{TL}\right)$$

It turns out that the BP temperature for this mixture is 92.1°C. Since the feed is at 92.0°C it is slightly subcooled and therefore all liquid. It might contribute more to the liquid flow below the feed stage because it could condense some of the vapor flowing into the feed stage. This all depends on the relationship between the feed enthalpy and the other streams as we shall see.

The unknowns and equations will be similar as we work our way down the column. If the EMO assumption was exactly correct, $L_0 = L_1$ and $V_1 = V_2$, but this is clearly not the case. Table 8.1 from Excel "Rigorous/10-Tray" shows the calculations after this first iteration. Note that the stage-by-stage calculation and the overall process mole balance don't agree. A new guess for x_D must be made. Try some of these calculations by hand.

For the final solution, let x_D be a variable and the difference (Δ) between the stage-by-stage calculations and the overall process mole balance be a constraint. Table 8.2 shows the final values.

TABLE 8.1

First Iteration of Ten-Tray Distillation Column Rigorous Analysis

Stage	L (mol/s)	x (mol B/mol)	V (mol/s)	y (mol B/mol)	T (°C)	Q (kJ/s)
TC	100.0	0.850			83.3	−4,766.4
1	97.3	0.691	150.0	0.850	87.0	0.00000
2	95.4	0.539	147.3	0.745	91.0	0.00000
3	94.4	0.424	145.4	0.646	94.4	0.00000
4	94.0	0.352	144.4	0.572	96.7	0.00000
5	191.8	0.312	144.0	0.525	98.1	0.00000
6	191.5	0.195	141.8	0.369	102.3	0.00000
7	192.1	0.101	141.5	0.211	106.1	0.00000
8	192.9	0.037	142.1	0.084	108.9	0.00000
9	193.7	−0.001	142.9	−0.002	110.7	0.00000
10	194.2	−0.022	143.7	−0.053	111.7	0.00000
Reboiler	50.0	−0.033	144.2	−0.082	112.2	4,835.1

OMB = 0.150

Δ = −0.183

TABLE 8.2

Final Iteration of 10-Tray Distillation Column Rigorous Analysis

Stage	L (mol/s)	x (mol B/mol)	V (mol/s)	y (mol B/mol)	T (°C)	Q (kJ/s)
TC	100.0	0.950			81.1	−4,668.9
1	98.5	0.880	150.0	0.950	82.6	0.00000
2	96.8	0.786	148.5	0.904	84.7	0.00000
3	95.1	0.677	146.8	0.842	87.3	0.00000
4	93.8	0.573	145.1	0.771	90.1	0.00000
5	191.9	0.489	143.8	0.704	92.4	0.00000
6	190.9	0.422	141.9	0.644	94.5	0.00000
7	189.9	0.337	140.9	0.554	97.2	0.00000
8	189.4	0.244	139.9	0.439	100.4	0.00000
9	189.5	0.160	139.4	0.314	103.7	0.00000
10	190.0	0.095	139.5	0.200	106.4	0.00000
Reboiler	50.0	0.050	140.0	0.111	108.3	4,728.9

OMB = 0.050

Δ = 0.000

The leaving and passing streams are indicated as stair steps on the xy graph in Figure 8.4. Also, the top operating line (TOL) and bottom operating line (BOL) using the McCabe–Thiele equations (Eqs. 7.4, 7.5, and 7.10) that use the EMO assumption are also drawn along with the stair steps for reference.

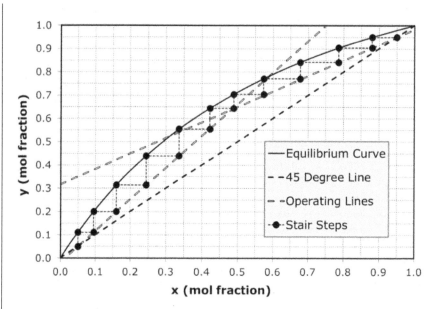

FIGURE 8.4

McCabe–Thiele diagram for Example 8.1 with the compositions for the stair steps from the rigorous calculations.

The rigorous compositions don't line up perfectly with the operating lines but the difference is small.

The effect of the EMO assumption is better seen with the flow profiles shown in Figure 8.5, the temperature profiles in Figure 8.6, and the composition profiles in Figure 8.7 obtained from the Lewis and rigorous methods. In this case, the vapor flow rate, for instance, increases from 140 to 150 mol/s as it rises through the column. Notice also that the liquid flow increases from 94 to 192 mol/s at the feed stage. As a slightly subcooled liquid, you might expect the liquid flow rate out of the feed stage to increase by more than 100 mol/s. Can you explain this? Go have a cup of coffee and study the Excel "Rigorous/10-Tray" results. I'll explain in the next section.

d. Explore the problem.

So why did the subcooled liquid feed contribute fewer than 100 mol/s to the liquid stream? It is only slightly subcooled; the BP is 92.1°C. The vapor coming into the feed stage from Tray 6 is at 94.5°C and the vapor leaves the feed stage at 92.4°C – both higher than the 92.0°C feed temperature. That vapor was hot enough, had enough energy, to raise the feed to saturation and then vaporize a small amount.

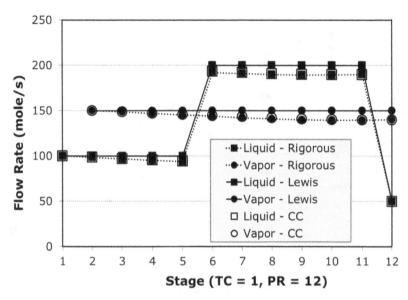

FIGURE 8.5
Molar flow rates (mol/s) of the vapor and liquid streams at each stage calculated by the Lewis method using the EMO assumption, using the adiabatic stage assumption and energy balances, and from the ChemCAD process simulator.

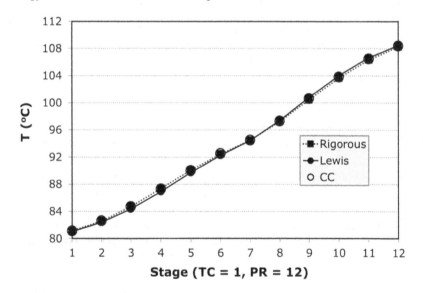

FIGURE 8.6
Temperature (°C) at each stage calculated by the Lewis method using the EMO assumption, using the adiabatic stage assumption and energy balances, and from the ChemCAD process simulator.

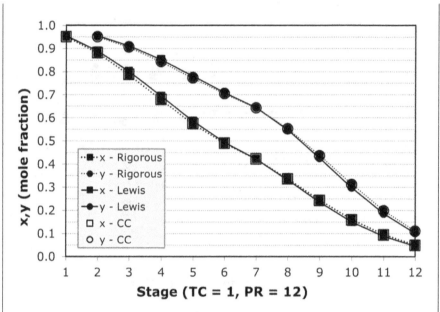

FIGURE 8.7
Benzene mole fraction of the vapor and liquid streams at each stage calculated by the Lewis method using the EMO assumption, using the adiabatic stage assumption and energy balances, and from the ChemCAD process simulator.

Study the results carefully. Make a list of any interesting observations or questions you might have about the results. Discuss with your study partners and ask tough questions to your TA and professor.

e. Be awesome!

Do you like Excel "Rigorous/10-Tray?" Why don't you hop into the driver's seat and give it a test ride? What happens when you change the temperature from a subcooled liquid to a mixed liquid/vapor to a superheated vapor? What happens if you change the feed location to a non-optimum tray? What happens to x_D as you increase or decrease R? Specify two of the variables R, D, or x_D and (slightly) modify Solver to iterate on the other. These are some of the *easy experiments* you can do! Use your process simulator (at UA we use ChemCAD but you might use Aspen or some other) to duplicate your experiments.

e⁺. Be really awesome!

No, that is not your grade ☺ Why not replace the benzene and toluene properties with those of another ideal pair of chemicals – say hexane and hexene? What kind of modifications would you need to make on the spreadsheet?

8.3 CAD Results

Figures 8.5–8.7 also show the results generated by the ChemCAD process simulator using the same process specifications. In the simulation, I chose Raoult's law for calculating K values and the program's thermodynamics wizard chose the non-random two-liquid (NRTL) model for calculating the enthalpies. The rigorous method and ChemCAD results track almost exactly.

8.4 Summary

The Lewis method (Chapter 6) was a stage-by-stage calculation using the EMO assumption. A more accurate analysis uses an enthalpy balance at each stage, assuming all trays are adiabatic. (Of course, this might not be valid either in a real process, so you would need other information.) This has the additional advantage that the heat added at the PR and removed at the TC is automatically calculated. The molar flow rates, compositions, and temperatures of each stage match those predicted by the commercial process simulator ChemCAD.

We will use the EMO assumption for most examples in Section II of this book. It is a bit tedious to include enthalpy calculations as we explore variations on distillation processes. Of course, you are ALWAYS welcome to modify solutions to include energy balances! It is just so quick and easy to compare your EMO-based results with a process simulator.

9

Distillation: Variations on a Theme

The classic distillation process consists of a column with packing or trays to promote liquid–vapor contact. A mixture of chemicals is fed at an optimum position. A vapor rich in the light components leaves the top and enters a total condenser (TC) where it is changed to a bubble point liquid. Part of that stream is recycled back to the top of the column as reflux and the rest is taken as distillate product. A liquid rich in the heavy components leaves the bottom and enters a partial reboiler (PR) where it is partially vaporized. The vapor is returned to the column as boilup and the liquid is taken as a bottoms product.

There must be a million variations (OK – I am probably exaggerating) on this process. You can handle every single one of them with the engineering tools you have already mastered. We'll just look at a few of the more common variations. Multicomponent and non-ideal systems will be addressed in Chapters 10 and 11, respectively.

9.1 Multiple Feeds

Two feed streams with the same components but different compositions might be more efficiently separated by feeding to different trays. As a very general rule, it is best to feed to a stage that is approximately at the feed composition. So, let's say stream #1 is 20% benzene and 80% toluene and stream #2 is 20% toluene and 80% benzene. The optimum feed tray for stream #1 will be toward the bottom of the column and for stream #2 will be toward the top of the column. We can find the specific optimum stage by using the graphical McCabe–Thiele method and then verify that with the analytical Lewis method.

EXAMPLE 9.1   Two Saturated Liquid Feed Streams

A distillation column is fed with two benzene/toluene saturated liquid streams. One is 100 mol/s, 80 mol% benzene, and the other is 100 mol/s, 20 mol% benzene. The column is equipped with a TC producing a bubble point distillate at a reflux ratio, $R = 1$, and a PR. The distillate product should be 95 mol% benzene and the bottom product should be

---

5 mol% benzene. Find the required number of equilibrium stages and the optimum feed stages. Assume equimolal overflow (EMO).

a. Draw a labeled process flow diagram.
   The process flow diagram (PFD) is shown in Figure 9.1.
b. Conduct a degree of freedom analysis.
   Overall process:

Unknowns (2): B and D

Equations (2):

   Total mole balance

   Benzene mole balance

With zero degrees of freedom (DOF), enough information is given to complete the overall mole balance.

c. Write the equations and solve.
   Total mole balance: $F_1 + F_2 = D + B$

   Benzene mole balance: $F_1 z_1 + F_2 z_2 = D x_D + B x_B$

   At last, a set of equations we can easily solve by hand! Both D and B will be 100 mol. Since $R = 1$, $L_0 = 100$ mol, and $V_1 = 200$ mol. At the top of the column we'll use flow symbols L and V, between the feeds we'll use L′ and V′, and at the bottom of the column we'll use L″ and V″. The top feed adds 100 mol of liquid so L′ = 200 mol. The bottom feed adds another 100 mol of liquid so L″ = 300 mol. A total mole balance around the reboiler shows that V″ = V′ = V = 200 mol since vapor is only generated by the reboiler. Add these values to Figure 9.1.

GRAPHICAL MCCABE–THIELE ANALYSIS

Print a blank xy graph from Excel "Variations/xy" and follow along. Using the McCabe–Thiele method, we can draw the top, middle, and bottom operating lines and then step off the stages, moving from one operating line to the next at the optimum stage. The operating lines switch at the feed lines where the slope, L/V, increases since we are adding a saturated liquid. The top feed line will be a vertical line (saturated liquid) starting at (0.8, 0.8) and the bottom feed line will also be a vertical line starting at (0.2, 0.2). Draw these lines on your xy graph.

   The top operating line (TOL) is

$$y = (L/V)x + (D/V)x_D = (100/200)x + (100/200)0.95 = 0.5x + 0.475$$

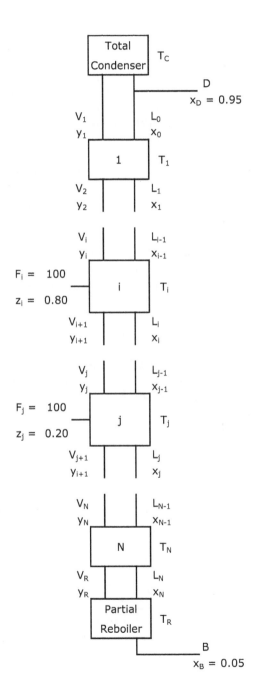

FIGURE 9.1
Process flow diagram of a two-liquid feed distillation column in Example 9.1.

The TOL includes the distillate composition, 0.95 mol B/mol, on the 45° line and the y-intercept will be (0, 0.475). Note that it intercepts the top feed line at (0.800, 0.875). This is where the middle operating line starts (between the two feeds). Draw the TOL on your xy graph.

We have a choice now to next draw the bottom operating line (BOL) or the middle operating line (MOL). Let's choose the BOL since we already have an equation for that.

$$y = (L''/V'')x - (B/V'')x_B = (300/200)x - (100/200)0.05 = 1.5x - 0.025$$

One point on this line is the bottoms composition, 0.05 mol B/mol, on the 45° line. We can draw a line from there with a slope of 1.5. Alternatively, we can just calculate another vapor composition at an arbitrary liquid composition. Might as well choose the bottom feed composition of 0.2 mol B/mol. That point is (0.20, 0.275). Go ahead and connect (0.050, 0.050) with (0.200, 0.275) on your xy graph. Note that the slope of this line is indeed 1.5.

$$BOL\ Slope = (0.275 - 0.05)/(0.2 - 0.05) = 1.5$$

The easiest way to draw the MOL at this point is to simply use the intersections that we already calculated for the TOL with the top feed line and the BOL with the bottom feed line. That is, we connect (0.200, 0.275) with (0.800, 0.875). Go ahead and connect those lines. The slope is 200/200 = 1 as expected.

$$MOL\ slope = (0.800 - 0.200)/(0.875 - 0.275) = 1$$

Alternatively, we could have started with the MOL, noting that one point is (0.800, 0.875) and the slope will be $L'/V' = 200/200 = 1$. The equation for the MOL will then be

$$Slope = 'rise/run' = (y - 0.875)/(x - 0.800) = 1$$

$$or\ y = x - 0.800 + 0.875 = x + 0.075$$

The y-intercept of this line is 0.075 so you can connect (0.800, 0.875) to (0, 0.075). The value of the MOL at the bottom feed composition is y = 0.200 + 0.075 = 0.275 as previously determined. Make sure your xy graph with the feed and operating lines looks like Figure 9.2.

Starting at the top, step off the stages, switching from the TOL to the MOL to the BOL at the optimum stages. Your final McCabe–Thiele graph should look like Figure 9.3. The separation takes just over 10

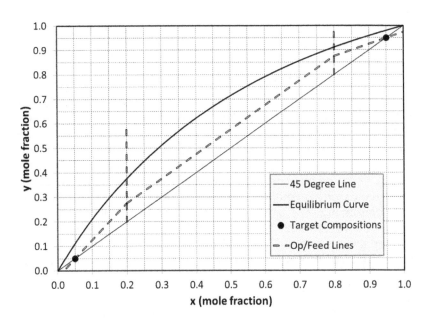

FIGURE 9.2
Operating and feed lines for Example 9.1.

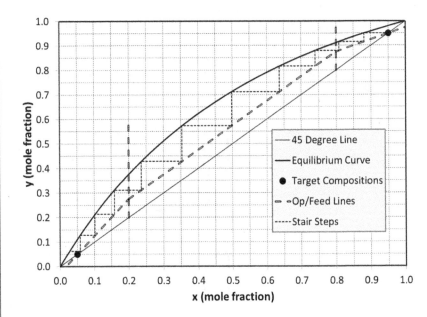

FIGURE 9.3
Completed McCabe–Thiele analysis of the distillation column in Example 9.1 with two liquid feeds.

stages (or R < 1), so we can use 10 trays and a PR. The optimum feed trays are Tray #3 and Tray #8.

LEWIS ANALYTICAL METHOD

In the Lewis method, we first set the number of stages and identify the feed trays. Using the results of the graphical McCabe–Thiele method, we'll specify 11 stages (ten equilibrium trays and a PR) and feed to Trays 3 and 8. We can set either the distillate composition or the reflux ratio and guess the other. Here, we will set $x_D = 0.95$ and *guess* the reflux ratio to be 1. Then we proceed with a stage-by-stage calculation involving mole balance, phase equilibrium, and the EMO assumption equations. At the bottom of the column we check to see if the bottoms composition matches the overall mole balance. If not, we go back and make another guess for the reflux ratio. If $x_D = 0.95$, the reflux ratio should turn out to be less than 1.0. The DOF analysis is identical to those we have done before.

a. Draw a labeled process flow diagram.

A generic PFD for any stage is drawn in Figure 9.4 to help generate the equations that need to be solved using the Lewis method. I'll note here that the same PFD can be used to analyze other situations, such as a partial condenser or a stage with a side product. Just eliminate missing streams.

Each equilibrium Stage i has a saturated vapor inlet stream, V_{i+1}, from the stage below, a saturated liquid inlet stream, L_{i-1}, from the stage above, a saturated liquid, L_i, and a saturated vapor, V_i, leaving the stage, a potential feed stream, F_i, and a potential

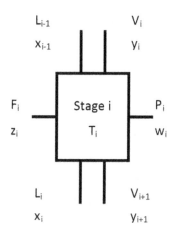

FIGURE 9.4
A single equilibrium stage with a feed and a product stream.

product stream, P_i. The feed stream could be a subcooled or saturated liquid, a mixture of vapor and liquid, or a saturated or superheated vapor depending on the process specifications. The side product will be a saturated liquid or saturated vapor at the tray composition, depending on where the stream is withdrawn. Mole fractions are denoted as always with x for liquids, y for vapors, z for feed, and w (new here) for product.

b. Conduct a degree of freedom analysis.

It might help to draw Figure 9.4 on a separate sheet of paper and only include the relevant streams for a specific stage.

TC

The only streams involved are the saturated vapor from Stage 1, V_1, and the saturated liquid product, L_C. x_D was specified to be 0.95 mol B/mol so all other compositions are the same. The reflux ratio was guessed to be R = 1.

Unknowns (3): V_1, L_C, T_C

Equations (3):

   1 total mole balance

   1 equilibrium relationship (benzene bubble point)

   1 process specification (reflux ratio and D solved by overall mole balance)

With zero DOF, the flow rates and temperature of the condenser can be calculated.

Stage i

For any stage, a liquid and vapor leave in equilibrium and a vapor enters from the stage below and a liquid enters from the stage above. All streams are saturated. A feed or product can enter or leave, but they will generally be completely specified. A product vapor will have composition y_i identical to the leaving vapor, and a product liquid will have composition x_i identical to the leaving liquid. The composition of the streams at the top of the stage (V_i and L_{i-1}) will have been calculated already for Stage i–1.

Unknowns (5): L_i, x_i, V_{i+1}, y_{i+1}, T_i

Equations (5):

   2 mole balances (total and benzene)

   2 equilibrium relationships (benzene and toluene or dew point)

   1 process specification (EMO)

With zero DOF, the unknowns for the stage can be calculated. In turn, the streams above the next stage are now known and the process is repeated stage by stage.

c. Write the equations and solve.

Without the above McCabe–Thiele analysis, we would not know which stages to feed. We would have to try different combinations to find the optimum stages by trial and error. However, we know that 3 and 8 are the feed stages, so $F_3 = 100$, $z_3 = 0.8$, $F_8 = 100$, and $z_8 = 0.2$. All other F_i and P_i are zero.

Condenser

Starting with the TC, we already specified that $x_D = 0.95$ and found from an overall balance that $D = 100$. If we guess $R = 1$, then $L_0 = 100$ and $V_1 = 200$. T_C is found to be 81.1°C using the bubble point equation for this composition. Also, $y_1 = x_0 = x_D$.

Stage 1

Total mole balance: $V_2 + L_0 = V_1 + L_1$

Benzene mole balance: $V_2 y_2 + L_0 x_0 = V_1 y_1 + L_1 x_1$

Benzene equilibrium: $y_1 = K_{1B} x_1$

Dew point: $y_1/K_{1B} + (1-y_1)/K_{1T} = 1$

EMO: $V_1 = V_2$

We know at this point L_0, x_0, V_1, and y_1. This leaves L_1, x_1, V_2, y_2, and T_1 to calculate with the five equations. The solution appears in Excel "Variations/Multiple Liquid Feeds" and is shown in the table below. Try doing this by hand. First calculate T_1 using y_1 with the dew point equation and the Excel tool "Goal Seek" (it is an iterative calculation). Then find x_1 using Raoult's law (you have K_{1B} from the first step). Use the EMO assumption to find $V_2 = V_1$. Then use the total mole balance to find L_1 (it will be the same as L_0) and the benzene mole balance to find y_2. Did you get the same values?

Stage	L (mol/s)	x (mol B/mol)	V (mol/s)	y (mol B/mol)	T (°C)
TC	100	0.950			81.1
1	100	0.881	200	0.950	82.6
2			200	0.915	

Repeat this procedure for Stage 2. Stage 3 has the feed stream so the equations are a little different.

Stage 3

Total mole balance: $V_4 + L_2 + F_3 = V_3 + L_3$

Benzene mole balance: $V_4y_4 + L_2x_2 + F_3z_3 = V_3y_3 + L_3x_3$

Benzene equilibrium: $y_3 = K_{3B}x_3$

Dew point: $y_3/K_{3B} + (1-y_3)/K_{3T} = 1$

EMO: $V_3 = V_4$

We know at this point F_3, z_3, L_0, x_0, V_1, and y_1. This leaves L_3, x_3, V_4, y_4, and T_3 to calculate with the five equations. The result to this stage is shown in the table below with previous results and specifications shaded. This process is repeated all the way down the column until we reach the reboiler.

Stage	L (mol/s)	x (mol B/mol)	V (mol/s)	y (mol B/mol)	T (°C)
TC	100	0.950			81.1
1	100	0.881	200	0.950	82.6
2	100	0.808	200	0.915	84.2
3	200	0.741	200	0.879	85.8
4			200	0.816	

Reboiler

For the reboiler there is no vapor from a stage below. We already know the streams above the reboiler (L_{10} and V_R). The overall mole balances were already solved to yield B.

Benzene equilibrium: $y_R = K_{RB}x_B$

Dew point: $y_R/K_{RB} + (1-y_R)/K_{RT} = 1$

We know at this point y_R. This leaves x_B and T_R to calculate with the two equations. The final solution for the guess of $R = 1$ is shown below.

This is exactly the solution from the McCabe–Thiele analysis. Take a colored pen to Figure 9.3 and plot the passing and leaving streams to prove this to yourself.

Stage	L (mol/s)	x mol B/mol	V (mol/s)	y (mol B/mol)	T (°C)
TC	100	0.950			81.1
1	100	0.881	200	0.950	82.6
2	100	0.808	200	0.915	84.2

(Continued)

Stage	L (mol/s)	x mol B/mol)	V (mol/s)	y (mol B/mol)	T (°C)
3	200	0.741	200	0.879	85.8
4	200	0.637	200	0.816	88.4
5	200	0.498	200	0.712	92.2
6	200	0.353	200	0.573	96.7
7	200	0.236	200	0.428	100.7
8	300	0.159	200	0.311	103.7
9	300	0.102	200	0.213	106.0
10	300	0.058	200	0.128	107.9
PR	100	0.028	200	0.063	109.3
	X_R by OMB	0.050			
	$\Delta =$	−0.022			

An overall benzene mole balance shows that the bottoms composition should be $x_B = 0.05$. So the guess of R = 1 was too high. Add the difference between the bottoms composition from a stage-to-stage analysis and from the overall mole balance as a constraint and let R vary. This change was made in Excel "Variations/Multiple Liquid Feeds" and the final result is shown in the table below and in Figure 9.5. It appears that R = 0.82 is sufficient to achieve 100 mol of distillate with 0.95 mol B/mol using 10 trays, a TC, a PR, and the feeds sent to Stages 3 and 8.

Stage	L (mol/s)	x (mol B/mol)	V (mol/s)	y (mol B/mol)	T (°C)
TC	82	0.950			81.1
1	82	0.881	182	0.950	82.6
2	82	0.815	182	0.919	84.0
3	182	0.759	182	0.889	85.3
4	182	0.676	182	0.841	87.4
5	182	0.557	182	0.759	90.5
6	182	0.417	182	0.639	94.6
7	182	0.291	182	0.500	98.8
8	282	0.198	182	0.373	102.2
9	282	0.140	182	0.279	104.5
10	282	0.089	182	0.189	106.6
PR	100	0.050	182	0.111	108.3
	X_R by OMB	0.050			
	$\Delta =$	−0.022			

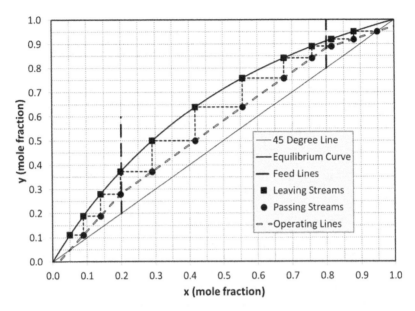

FIGURE 9.5
Completed Lewis method solution for Example 9.1 with two liquid feeds.

    d. Explore the problem.

        Try feeding to other stages and compare performance to these optimum feed stages. Note that the composition of Stage 8 falls almost exactly on the bottom feed line, so feeding to Stage 9 shouldn't make much of a difference.

    e. Be awesome!

        Try inputting all of these equations into Excel "Variations/MLF DIY." Make sure you achieve identical results.

TRY THIS AT HOME 9.1   Multiple Vapor Feeds

The same specifications as Example 9.1 are given except that the feed streams are saturated vapors. A distillation column is fed with two benzene/toluene saturated vapor streams, each 100 mol/s. One is 80 mol% benzene and the other is 20 mol% benzene. The column is equipped with a TC producing a bubble point distillate at a reflux ratio, $R = 1$, and a PR. The distillate product should be 95 mol% benzene and the bottom product should be 5 mol% benzene. Find the required number of stages and the optimum feed stages.

a. Draw a labeled process flow diagram.

   The overall PFD is exactly the same as Figure 9.1. The only difference is the fact that the feed streams are saturated vapors, not saturated liquids.

b. Conduct a degree of freedom analysis.

   For the overall process we might be able to calculate stream flow rates but we already know that compositions must be found by a stage-to-stage calculation. The DOF was done in Example 9.1.

c. Write the equations and solve.

   The EMO assumption and reflux ratio process specification can be used to find intermediate flow rates. Since $R = 1$, $L_0 = L$ (top of column) = 100 mol/s. There are no liquid feeds so $L = L' = L'' = 100$ mol/s. $V_1 = D + L_0 = 200$ mol/s. Above the top feed stage then, $V = V_1 = 200$ mol/s. Below the top feed stage the vapor flow rate will be $V' = V - F_1 = 100$ mol/s. Below the bottom feed stage $V'' = V' - F_2 = 0$ mol/s. This means that none of the liquid leaving the bottom of the column is vaporized as boilup; it all leaves as the bottoms product. Since there is no vapor–liquid contact below the bottom feed stage, no separation occurs there – the section is useless!

   Let's look at this on a McCabe–Thiele graph. Print an xy graph from Excel "Variations/xy" and mark the top and bottom compositions on the 45° line at $x_D = 0.95$ and $x_B = 0.05$. Recall that the feed lines will be horizontal for saturated vapor feeds. Locate those compositions, $z_1 = 0.80$ and $z_2 = 0.20$, on the 45° line and draw horizontal lines to the left. Now, draw the TOL from x_D to the y-intercept at $x_D/(1 + R) = 0.95/2 = 0.475$. Notice there is already a problem appearing; the number of stages above the top feed stage is going to be large.

   From the intersection of the TOL and top feed line draw a line with slope $L'/V' = 1$ all the way to the intersection with the y-axis. Notice that it intersects the equilibrium line before it intersects the bottom feed line. This is a type of pinch point that prevents the specified separation even with an infinite number of stages.

   Even worse, starting at x_B on the 45° line, the BOL has a slope of $-L''/V'' = -100/0 = \infty$. That certainly won't work! We have to find conditions that increase the slope of the MOL and decrease the slope of the BOL so that they can intersect the bottom feed line below the equilibrium curve. Take a moment to think about that and compare your graph to Figure 9.6.

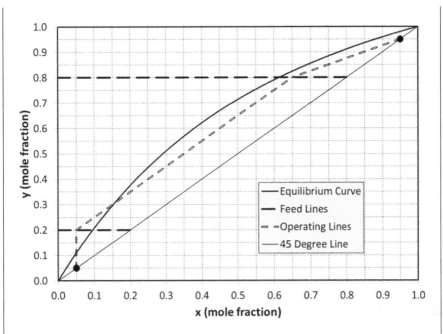

FIGURE 9.6
Operating lines for a distillation column with two vapor feeds and $R = 1$ according to Try This at Home 9.1.

Notice that a pinch point occurs at around (0.15, 0.30) if you try stepping off stages from the top. Operating lines must stay below the equilibrium curve or the separation is impossible to achieve at these initial specifications. Does this mean the column would blow up or launch like a rocket into orbit?! No, it just means that you could not achieve the specified distillate and bottoms compositions. Adding more stages will obviously not help. You must either increase the reflux ratio or reduce the target compositions.

Now you can see an important benefit of the McCabe–Thiele method. It allows us to visualize the process and to spot and understand problematic or impossible specifications. An algebraic solution in Excel would probably just give us an error message and it might be difficult to find the error.

TRY THIS AT HOME 9.1   Multiple Vapor Feeds (Revised)

Using $R = 1$ made it impossible to achieve the specifications of the original problem. Let's try doubling the reflux ratio. A distillation column is fed with two benzene/toluene saturated vapor streams, each 100 mol/s. One is 80 mol% benzene and the other is 20 mol% benzene. The column is equipped with a TC producing a bubble point distillate at a reflux ratio, $R = 2$, and a PR. The distillate product should be 95 mol% benzene and the bottom product should be 5 mol% benzene. Find the required number of stages and the optimum feed stages.

a. Draw a labeled process flow diagram and

b. Conduct a degree of freedom analysis

Since reflux ratio was not indicated on the PFD for Example 9.1, we can continue to use Figure 9.1. Use the reflux ratio and EMO assumption to find all other vapor and liquid stream flow rates. As before, $D = B = 100$ from an overall mole balance. $L = L' = L'' = L_0 = R D = 2 \times 100 = 200$, $V = L + D = 200 + 100 = 300$, $V' = V - F$ (top) $= 300 - 100 = 200$, and $V'' = V' - F$ (bottom) $= 200 - 100 = 100$.

c. Write the equations and solve.

McCabe–Thiele (MT) Solution
The MT solution is proving to be a quick screening tool to see if a problem specification is even possible. Print a copy of the xy graph from Excel "Variations/xy." Locate the distillate and bottom compositions and the feed line intersections on the 45° line. Draw the TOL from the distillate composition to the y-intercept and the BOL from the bottoms composition with a slope L''/V''.

TOL: $y = (L/V)x + (D/V)x_D = (100/300)x + (100/300)0.95$

$\qquad = 1/3x + 0.317$

BOL: $y = (L''/V'')x - (B/V'')x_B = (100/100)x - (100/100)0.05$

$\qquad = x - 0.05$

Notice that they intersect the feed lines below the equilibrium line so this separation is possible. Draw the MOL between the two feed lines. Now you are ready to step off the stages to find the number of stages and the optimum feed stages. There are ten trays and a PR. The optimum feed trays are 3 and 8 or 9. Note that it doesn't change much

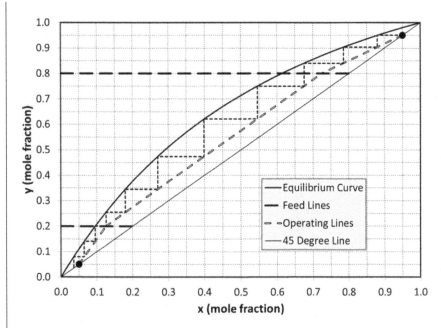

FIGURE 9.7
Completed McCabe–Thiele diagram for two vapor feeds according to Try This at Home: 9.1.

if you step off from the MOL to the BOL at either stage. Compare your solution to Figure 9.7.

Lewis Solution
Use Excel "Variations/Multiple Vapor Feeds" with a reflux ratio R = 2 to perform the stage-by-stage calculations. Your result should look exactly like the McCabe–Thiele solution with a concentration of benzene in the bottoms stream less than specified. Now, in Solver set the difference between the bottoms composition from the stage-by-stage calculation and overall mole balance (Δ) to zero and make the appropriate changes to vary the reflux ratio. Your final result will look like Figure 9.8. The required reflux ratio is now only 1.87.

d. Explore the problem.
   Move the feed streams to non-optimal stages and observe how that changes the composition of the products. Change the flow rate and composition of the feed streams. Perform a quick McCabe–Thiele graphical analysis first to make sure your new specifications are actually possible.

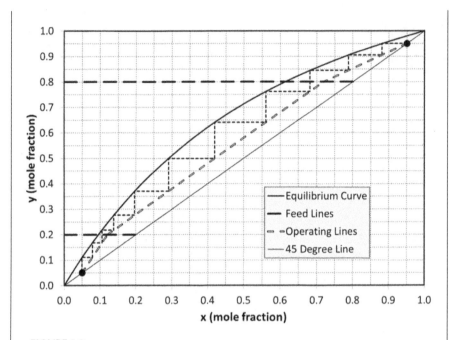

FIGURE 9.8
Completed Lewis method solution for two liquid feeds according to Try This at Home 9.1.

    e. Be awesome!
        Create your own process simulation using Excel "Variations/MVF DIY." Make sure you calculate identical results. Be even more awesome by adding/removing stages to see how that affects the column performance.

9.2 Open Steam

Instead of using a reboiler, heat can be added to the bottom of a column by direct injection of saturated steam. This will make the most sense for separations involving water as the heavy component. The example will use separation of water/ethanol and just the graphical technique to avoid (for the moment) using liquid-phase activity coefficients for this non-ideal system.

    The major difference between this and the classic distillation system occurs at the bottom of the column, which contains no reboiler. Liquid enters the bottom section from above and is immediately mixed with saturated steam. The bottoms stream leaving this last stage passes the steam line for a pair of compositions located on the x-axis (y = 0, the vapor composition). This will be one end of the BOL. The rest of the McCabe–Thiele analysis remains the same.

Operating Lines

The top operating line will remain

$$y = (L/V)x + (D/V)x_D \tag{7.4}$$

The bottom of the column is shown in Figure 9.9. The flow rate of steam is S (mol W/s) and has no ethanol. The total mole balance around the bottom stage is

$$L' + S = V' + B \tag{9.1}$$

and the ethanol mole balance is

$$L'x_{N-1} = V'y_N + Bx_B \tag{9.2}$$

where V′ and L′ are the vapor and liquid flow rates below the feed stage. In a straight-line form, the BOL is then

$$y = (L'/V')x - (B/V')x_B \tag{9.3}$$

where x and y are understood to be passing streams below the feed stage. This is the same equation for the BOL as before because the open steam does not contribute to the ethanol mole balance. The open steam and the bottoms are passing streams, so does (x, y) = (x_B, 0) lie on the BOL? Substituting these values into Equation 9.3

$$0 = (L'/V')x_B - (B/V')x_B \text{ or } L' = B$$

and the EMO assumption requires this to be true.

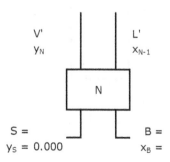

FIGURE 9.9
Bottom of a distillation column with steam injection instead of a reboiler.

EXAMPLE 9.3   Open Steam Distillation of Ethanol and Water

A total of 100 mol/s of 20 mol% ethanol in water enters a distillation column with a TC and open steam injection to the bottom tray. The feed is
a mixture that splits evenly between the liquid and vapor streams in the
column. The reflux ratio is 2. The bottom product is a 99.0 mol% water
mixture and the distillate is 75.0 mol% ethanol. Determine the number
of required stages, the optimum feed tray, and all flow rates and compositions. Assume perfectly efficient stages and equimolal overflow.

a. Draw a labeled process flow diagram.
   The labeled overall PFD is shown in Figure 9.10. This is sufficient to set up the graphical analysis.
b. Conduct a degree of freedom analysis.

   Unknowns (3): S, D, B

   Equations (3):
      2 mole balances (total and ethanol)
      1 process specification (reflux ratio)

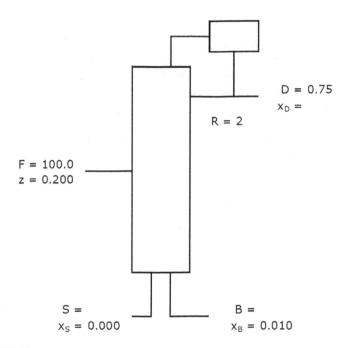

FIGURE 9.10
PFD of an open steam distillation column according to Example 9.3.

c. Write the equations and solve.

Total mole balance: $S + F = D + B$

Ethanol mole balance: $Fz = Dx_D + Bx_B$

Using the reflux ratio is a bit tricky. We know from the EMO assumption and the process description that $V_1 = S + \frac{1}{2}F = D + L = D + 2D = 3D$ because half the feed enters into the vapor stream.

Process specification: $S + \frac{1}{2}F = 3D$

These are solved in Excel "Open Steam/Operating Lines" and the values are $D = 25.3$, $B = 100.6$, and $S = 26.0$ mol/s. Assuming EMO, $V = 76.0$, $L = 50.6$, $V' = 26.0$, and $L' = 100.6$. Substituting these values into the appropriate equations, we find that

$$\text{TOL: } y = 0.667x + 0.25$$

$$\text{BOL: } y = 3.875x - 0.039$$

$$\text{Feed line: } y = -x + 0.40$$

Be sure to develop these on your own. The three lines are plotted in Figure 9.11. They all intersect at (0.090, 0.310). In fact, a quick way of drawing the BOL is to start at the bottom passing streams, $(x_B, y_S) = (0.010, 0.000)$, and draw a straight line to the TOL-feed line intersection.

The graph is now ready to begin stepping off the stages. It doesn't matter if you start at the top or bottom. In Figure 9.12, I started at the bottom because the analytical solution (Excel "Open Steam/Lewis") requires the activity coefficient that in turn depends on the x value. The vapor leaving the bottom stage must be in equilibrium with the liquid leaving the stage. So start by drawing a vertical line from (0.010, 0.000) to the equilibrium curve (0.010, 0.111). The liquid passing stream will be on the BOL at $y = 0.111$, or in this case $x = 0.039$.

This process is repeated until the third stage from the bottom, where bigger steps can be achieved by switching to the top operating line. This will be the feed stage. Finally, at the

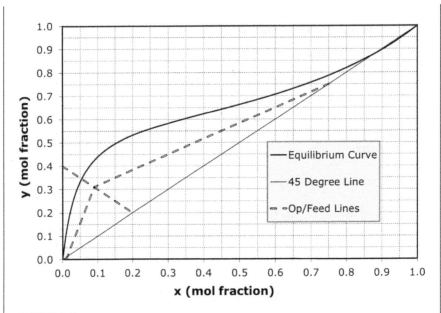

FIGURE 9.11
Top and bottom operating lines and the feed line. Note that the passing streams at the bottom of the column lie on the x-axis.

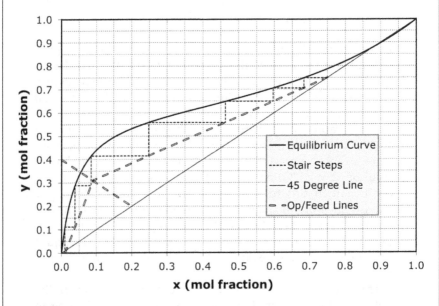

FIGURE 9.12
McCabe–Thiele solution for the open steam distillation column specified in Example 9.3.

seventh stage from the bottom, the target distillate composition is almost achieved. It is so close that we could try increasing the reflux ratio just a hair.

d. Explore the problem.

Vary the reflux ratio to see the effect on the number of stages and optimum feed tray. Try increasing it to 5 in Excel "Open Steam/MT" and printing Excel "Open Steam/Op Lines" after running Solver. You should find five stages with the feed going to Stage 4 (from the top) will almost reach the desired distillate composition. This will introduce some computational difficulties on Excel "Open Steam/Lewis" if you try to do this analytically. This is "easily" solved by changing Solver to act on fewer stages and changing the feed stream location.

Is there a minimum number of stages at an infinite reflux ratio? Print a blank xy graph from Excel "Open Steam/xy" and perform that calculation. Give the simulation a really large reflux ratio like 50 and see what happens. Compare to Figure 9.13.

Now try a smaller reflux ratio of 1.45. Print a blank xy graph and add the operating lines. Compare this to Excel

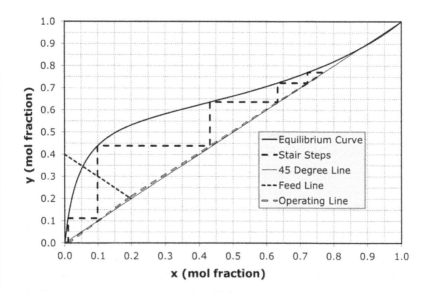

FIGURE 9.13

McCabe–Thiele analysis of the open steam distillation column from Example 9.3 with $R = 50$, essentially infinite reflux, which yields the minimum number of stages to achieve the specifications.

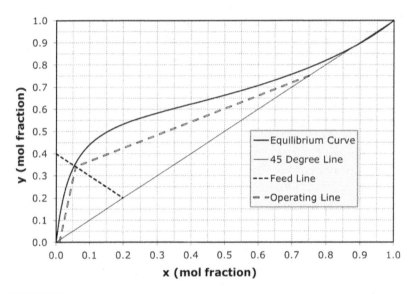

FIGURE 9.14
Minimum reflux ratio for the open steam distillation column specified in Example 9.3. Notice the pinch point where the feed line, operating lines, and the equilibrium curve intersect.

"Open Steam/Op Lines" after making the required change on Excel "Open Steam/MT" and running Solver. Do you see a pinch point developing? How many stages are now required and what is the optimum feed stage? Compare to Figure 9.14.

e. Be awesome!
   Make your own simulation using the Lewis method and Excel "Open Steam/DIY." Make sure you can successfully reproduce the above examples.

9.3 Side Products

It is common to take a side stream from the distillation column. This is especially true if you have a multicomponent distillation and want a specific product that is concentrated in the middle of the column. Sometimes this stream is even cooled or heated and put back into the column to improve operation (pump around). Let's keep it simple here and just take a single side stream from the benzene–toluene distillation column. It is best to demonstrate this with a specific example.

EXAMPLE 9.4   Liquid Side Stream Above Feed Stage

A distillation column is fed with 100 mol/s of a 30 mol% benzene/70 mol% toluene saturated liquid stream. The column is equipped with a TC producing a bubble point distillate at a reflux ratio, R = 3, and a PR. About 20 mol/s of distillate is produced with 95.5 mol% benzene and 5 mol/s of a saturated liquid is taken from Tray 4, which is above the optimal feed stage. Find the required number of stages, the optimum feed stage, and all stream flow rates, compositions, and temperatures.

a. Draw a labeled process flow diagram.

Figure 9.15 is a PFD of the column. The side stream is shown leaving Tray 4. We don't yet know the optimal feed tray so this is labeled Tray i. The bottom tray is unknown so it is labeled Tray N. This will be sufficient to write the overall mole balances.

b. Conduct a degree of freedom analysis.

Total condenser

Unknowns (4): L_0, T_C, V_1

Equations (4):

    1 mole balance (B)

    1 process specification (R)

    1 equilibrium relationship (dew point for T_C)

Because we specified D, R, and x_D (didn't have to guess anything because we will be determining the number of stages required), we have zero DOF and this part is solvable.

Tray 1

Unknowns (5): V_2, y_2, T_2, L_1, x_1

Equations (5):

    2 mole balances (total and B)

    2 equilibrium relationships (B and T)

    1 process specification (EMO)

We know the properties of L_0 and V_1 already from the condenser calculations, so there is zero DOF. We can continue down the column like this until something changes – like the side stream tray.

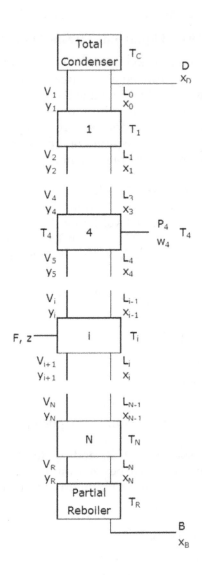

FIGURE 9.15
PFD of a distillation column with a side stream above the feed stage as described in
Example 9.4.

Tray 4

Unknowns (6): V_5, y_5, L_4, x_4, T_4, w_4

Equations (6):

 2 mole balances (total and B)

 2 equilibrium relationships (B and T)

1 process specification (EMO)

1 process specification (both leaving liquid streams have same composition)

The last equation is more like an assumption. Anyway, it looks like zero DOF. The rest of the column is analyzed like a standard column and we know the problem can be solved.

For a graphical solution, we just draw the operating lines and step off as many stages as needed. A slight complication is that we don't know the liquid composition for starting the MOL, but I'll show you that soon. For an analytical solution, we must set the number of stages, feed stage, and side stream stage, and guess x_D or R because we can't really guarantee a specific separation can take place.

c. Write the equations and solve.

Graphical McCabe–Thiele Method
We have enough information to draw the TOL but not where to start the MOL until we actually step off the stages to Tray 4.

$$L_0 = RD = (3)(20) = 60 \text{ mol/s}$$

$$V_1 = L_0 + D = 60 + 20 = 80$$

The TOL equation is

$$y = \left(\frac{L}{V}\right)x + \left(\frac{D}{V}\right)x_D = \left(\frac{60}{80}\right)x + \left(\frac{20}{80}\right)0.955 = 0.75x + 0.24$$

The TOL is drawn in Figure 9.16 with the stages stepped off until Tray 4. The liquid and vapor composition $(x_4, y_5) = (0.53, 0.64)$ becomes the starting point for the MOL. The slope of the MOL is the new $L/V = (60-5)/80 = 0.69$ and the equation is from rise/run.

$$\frac{(0.64 - y)}{(0.53 - x)} = 0.69$$

and the y-intercept (where x = 0) is

$$y_{int} = -(0.69)(0.53) + 0.64 = 0.27$$

The MOL is added in Figure 9.17 and the stages stepped off until they pass the feed composition z = 0.3.

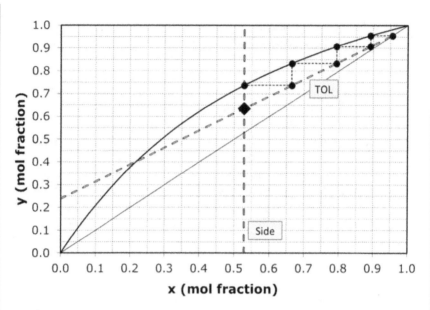

FIGURE 9.16
McCabe–Thiele diagram with TOL and stages up to Tray 4 identified according to Example 9.4.

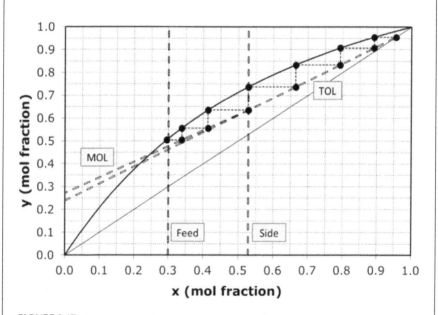

FIGURE 9.17
McCabe–Thiele diagram with BOL and stages up to Tray 7 identified according to Example 9.4.

The BOL will be drawn from (x_B, x_B) to the intersection of the MOL and feed line. x_B can be calculated (now that we know w_4!) from an overall mole balance.

$$Fz = Dx_D + Bx_B + P_4x_4$$

$$B = F - D - P = 100 - 20 - 5 = 75 \text{ mol/s}$$

$$x_B = (Fz - Dx_D - P_4x_4)/B = \{(100)/(0.3) - (20)(0.955) - (5)(0.53)\}/75 = 0.11$$

The BOL is added to Figure 9.18 and the stages stepped off until x_B is reached or exceeded. The stairs end exactly at $x_B = 0.11$ with ten equilibrium trays and a total reboiler. As we saw before, it will not necessarily "land" on an even number of stages, but I worked it out ahead of time analytically and used the results for the problem statement. Here, I'll show you!

Analytical Lewis Method
Let's specify ten equilibrium trays, a TC, and a PR. Feed 100 mol/s of 30 mol% benzene/70 mol% toluene to Tray 7 and remove 5 mol/s of liquid from Tray 4. We'll take 20 mol/s as distillate with a reflux ratio R = 3. Starting at the top of the column, recall that we must guess a distillate composition,

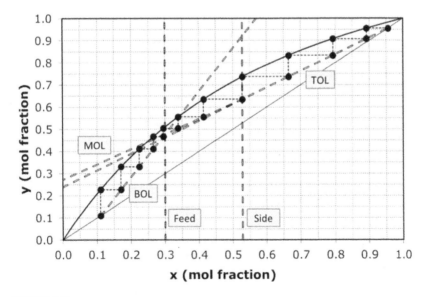

FIGURE 9.18
McCabe–Thiele diagram with BOL and all stages identified according to Example 9.4.

let's use $x_D = 0.900$, and correct it if the stage-by-stage analysis doesn't agree with the overall mole balance. Of course, we already know it doesn't.

TC/Split Point

Total mole balance: $V_1 = D + L_0$

Dew point: $y_1/K_{B1} + (1-y_1)/K_{T1} = 1$ (To find T_1)

Bubble point: $K_{BC}x_D + K_{TC}(1-x_D) = 1$ (To find T_C)

Process specification (reflux ratio): $L_0 = DR$

We already know D, R, and guessed x_D. The equations are solved in Excel "Side Stream/Lewis" and the final results for L_0, V_1, and T_C are shown here. The shaded cells have values that were specified.

Stage	F (mol/s)	z (mol B/mol)	L (mol/s)	x (mol B/mol)	V (mol/s)	y (mol B/mol)	T (°C)
TC			60	0.900			82.2
1					80	0.900	

Tray 1

Total mole balance: $L_0 + V_2 = L_1 + V_1$

Benzene mole balance: $L_0x_0 + V_2y_2 = L_1x_1 + V_1y_1$

Benzene equilibrium: $y_1 = K_{B1}x_1$

Dew point equation: $y_1/K_{B1} + (1-y_1)/K_{T1} = 1$ (To find T_1)

Process specification (EMO): $L_1 = L_0$ (Or use $V_2 = V_1$)

We already know the flow rates, compositions, and temperature of the streams at the top of Tray 1 (V_1, L_0) from the condenser/split point calculation. The solutions for L_1, x_1, V_2, y_2, and T_1 are shown here with the values previously calculated shaded. The calculations are continued this way until we reach Tray 4 with the side stream.

Stage	F (mol/s)	z (mol B/mol)	L (mol/s)	x (mol B/mol)	V (mol/s)	y (mol B/mol)	T (°C)
TC			60	0.900			82.2
1			60	0.779	80	0.900	84.9
2					80	0.809	

Tray 4

Total mole balance: $L_3 + V_5 = L_4 + V_4 + P_4$

Benzene mole balance: $L_3 x_3 + V_5 y_5 = L_4 x_4 + V_4 y_4 + P_4 w_4$

Benzene equilibrium: $y_4 = K_{B4} x_4$

Dew point equation: $y_4/K_{B4} + (1-y_4)/K_{T4} = 1$ (to find T_4)

Process specification (EMO): $L_4 = L_3$ (or use $V_5 = V_4$)

Process specification (perfectly mixed liquid phase): $w_4 = x_4$

We already know the flow rates, compositions, and temperature of the streams at the top of Tray 4 (V_4, L_3) from the Tray 3 calculation. The solutions for L_4, x_4, V_5, y_5, and T_4 are shown here with the values previously calculated or specified shaded. The calculations are continued this way until we reach Tray 7 with the feed stream.

Stage	F (mol/s)	z (mol B/mol)	L (mol/s)	x (mol B/mol)	V (mol/s)	y (mol B/mol)	T (°C)
TC			60	0.900			82.2
1			60	0.779	80	0.900	84.9
2			60	0.627	80	0.809	88.6
3			60	0.479	80	0.695	92.7
4	−5	0.364	55	0.364	80	0.584	96.3
5					80	0.498	

Tray 7

Total mole balance: $L_6 + V_8 + F = L_7 + V_7$

Benzene mole balance: $L_6 x_6 + V_8 y_8 + Fz = L_7 x_7 + V_7 y_7$

Benzene equilibrium: $y_7 = K_{B7} x_7$

Dew point equation: $y_7/K_{B7} + (1-y_7)/K_{T7} = 1$ (To find T_7)

Process specification (EMO): $L_7 = L_6 + F$ (Or use $V_8 = V_7$)

We already know the flow rates, compositions, and temperature of the streams at the top of Tray 7 (V_7, L_6) from the Tray 6 calculation. Also, the feed stream was completely specified. The solutions for L_7, x_7, V_8, y_8, and T_7 are shown here with the values previously calculated or specified shaded. The calculations are continued this way until we reach the PR.

Stage	F (mol/s)	z (mol B/mol)	L (mol/s)	x (mol B/mol)	V (mol/s)	y (mol B/mol)	T (°C)
TC			60	0.900			82.2
1			60	0.779	80	0.900	84.9
2			60	0.627	80	0.809	88.6
3			60	0.479	80	0.695	92.7
4	–5	0.364	55	0.364	80	0.584	96.3
5			55	0.289	80	0.498	98.8
6			55	0.250	80	0.446	100.3
7	100	0.300	155	0.230	80	0.419	101.0
8					80	0.318	

PR

The results at the PR are shown here with shading used to visually separate rows. Note that x_R by an overall mole balance does not agree with x_R from a stage-by-stage calculation.

Stage	F (mol/s)	z (mol B/mol)	L (mol/s)	x (mol B/mol)	V (mol/s)	y (mol B/mol)	T (°C)
TC			60	0.900			82.2
1			60	0.779	80	0.900	84.9
2			60	0.627	80	0.809	88.6
3			60	0.479	80	0.695	92.7
4	–5	0.364	55	0.364	80	0.584	96.3
5			55	0.289	80	0.498	98.8
6			55	0.250	80	0.446	100.3
7	100	0.300	155	0.230	80	0.419	101.0
8			155	0.163	80	0.318	103.6
9			155	0.089	80	0.188	106.6
10			155	0.020	80	0.045	109.7
PR			75	–0.036	80	–0.089	112.4
			X_R by OMB	0.136			
			$\Delta =$	–0.172			

Stage	F (mol/s)	z (mol B/mol)	L (mol/s)	x (mol B/mol)	V (mol/s)	y (mol B/mol)	T (°C)
TC			60	0.955			81.0
1			60	0.892	80	0.955	82.3
2			60	0.793	80	0.907	84.5
3			60	0.664	80	0.834	87.7
4	−5	0.529	55	0.529	80	0.737	91.3
5			55	0.414	80	0.635	94.7
6			55	0.338	80	0.556	97.2
7	100	0.300	155	0.295	80	0.504	98.7
8			155	0.265	80	0.467	99.7
9			155	0.224	80	0.411	101.2
10			155	0.171	80	0.331	103.2
PR			75	0.110	80	0.228	105.7
			X_R by OMB	0.110			
			$\Delta =$	0.000			

In Excel "Side Stream/Lewis," modify Solver to make Δ (the difference once again between x_R by an overall mole balance and x_R from the stage-by-stage Lewis method) zero by varying x_D. Solver will repeat the above calculations with new values of x_D until that happens. The final results are shown above.

The leaving and passing streams could be plotted as a McCabe–Thiele diagram and the results would look exactly like Figure 9.18.

d. Explore the problem.

Make small changes in the process specifications and any modifications needed in Solver and try to explain the results to your study partners. Specific suggestions are in Excel "Side Stream/Lewis." The McCabe–Thiele diagrams will be automatically updated. Make your own hand drawn McCabe–Thiele diagram for any new specifications and compare with Excel "Side Stream/MT Diagram."

e. Be awesome!

Use Excel "Side Stream/DIY" to input equations and compare solutions with Excel "Side Stream/Lewis."

e+. Be even more awesome!

Can you dream up other variations? How about subcooled reflux, subcooled feed, non-adiabatic stages, etc.? Some of these involve energy balances, but you are becoming awesome and can handle that.

9.4 Summary

Several variations of the classical distillation problem were examined. These included multiple liquid and vapor streams, open steam, and a side stream product. No new equations or concepts were required. The analysis simply required applying the tools you already have to the stages with new complications such as the feed tray, the reboiler (or lack of!), and the side stream tray.

10

Multicomponent Distillation

Two-component systems like benzene/toluene or ethanol/water give us simple cases to introduce distillation processes and provide insight into their operation. However, few distillation columns will be used to separate just two components. For instance, benzene and toluene streams frequently contain significant amounts of xylenes, ethyl benzene, and cumene. Ethanol and water streams might also contain methanol and other alcohols. Adding more components doesn't change the fundamental analysis of a column but it does add some complexity with more, but not different, equations. Also, graphical methods like McCabe–Thiele no longer work in two dimensions. So we are limited to analytical solutions or approximate methods (discussed in Seader et al., 2015).

10.1 Lewis Method

Look at a system containing three components: A, B, and C. Consider first the Lewis method in which we solve for unknown process variables at the top of the column and work our way down. This involves some initial guesses and iteration. A mole balance envelope is drawn around the condenser and split point in Figure 10.1. The streams crossing this boundary are the distillate, vapor from Stage 1, and the reflux to Stage 1. The unknown variables are the flow rates, compositions, and temperature. Since there is no separation in the total condenser (TC), the compositions of V_1 and L_0 are identical to the distillate. We can perform a degree of freedom analysis.

Unknowns (7): D, x_{DA}, x_{DB}, x_{DC}, V_1, L_0, T_C

> These process variables are related by mole balances (but just the total), equilibrium relationships (just the bubble point equation), summations, and various process specifications. Of course, we could just use two mole fractions, noting that they sum to one.

Equations (3):

    1 mole balance (total)

    1 equilibrium relationship (bubble point to find T_C)

    1 summation (for the distillate stream)

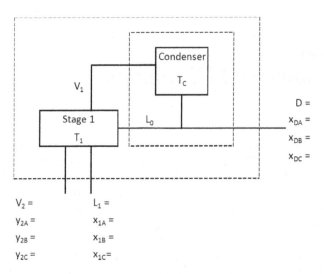

FIGURE 10.1
Process flow diagram of the top of a distillation column that separates three components.

This leaves four DOFs. Typically, we specify the reflux ratio, $R = L_0/D$. Also, we can specify the amount of distillate, D. At this point for two-component distillation, we guessed or specified a distillate composition. Then we could calculate all unknowns sequentially down the column. At the bottom, we checked if the overall process mole balance was satisfied. If not, we made another guess at the distillate composition and repeated the calculation. For this three-component distillation, we simply have to guess two compositions, x_{DA} and x_{DB} and check both the A and B overall mole balances at the bottom of the column.

Another mole balance envelope is drawn around the top of the column in Figure 10.1. The streams crossing this boundary are D, V_2, and L_1. Since the distillate flow rate and composition are now known or guessed, the only unknowns are the flow rates and compositions of the streams passing between Stages 1 and 2 along with the temperature of Stage 1. We can perform another degree of freedom analysis.

Unknowns (9): V_2, y_{2A}, y_{2B}, y_{2C}, L_1, x_{1B}, x_{1B}, x_{1C}, T_1

Equations (9):

   3 equilibrium relationships (benzene, toluene, and o-xylene)

   3 mole balances (total, benzene, and toluene)

   2 summation equations (streams V_2 and L_1)

   1 process specification (EMO: $V_2 = V_1$ or $L_1 = L_0$)

There are zero DOF for the top of the column. The flow rates and compositions of all streams can be determined down the column. The bottoms stream compositions can be compared to an overall mole balance around the column. If they don't match, another guess can be made for x_{DA}, and x_{DB}, and the process is repeated.

Note that we could just as easily have drawn the mole balance envelope (MBE) around Stage 1 alone since we know V_1, L_0, and their compositions from the first calculation. In fact, that is how our spreadsheet solution will be constructed.

10.2 "Brute Force" Method

Another approach is to look at all unknown variables and equations in the whole process – a "brute force" method. For the three-tray column with TC and partial reboiler (PR) shown in Figure 10.2, assume the feed flow rate and composition, reflux ratio, and distillate flow rate are specified. Can you count 38 unknown process variables (temperatures, flow rates, and compositions) and come up with 38 equations that relate them (process specifications, mole balances, equilibrium relationships, and summations)? Try it!

Of course, other process specifications are frequently used instead of D and R, such as the boilup ratio (V_R/B), percent recovery of a species, or a specific composition. You must make a reasonable guess for all process variables and then let Solver use an efficient algorithm to solve for all 38 variables as done in Example 10.1.

EXAMPLE 10.1   Separation of BHO by Distillation in a Three-Tray Column

A mixture of alkanes, butane (B), hexane (H), and octane (O) should be an ideal solution. A total of 100 mol/s of a saturated liquid mixture with 40% butane, 20% hexane, and balance octane is fed to the second tray of a three-tray distillation column equipped with a TC and PR. The distillate flow rate is 60 mol/s and the reflux ratio is 2.0. Assume EMO and that equilibrium is reached at each stage. Find the flow rate and composition of all streams and the temperature at each stage.

a. Draw a labeled process flow diagram.
    The PFD was already shown in Figure 10.2. Pencil in the values for any process variables specified in the problem statement.

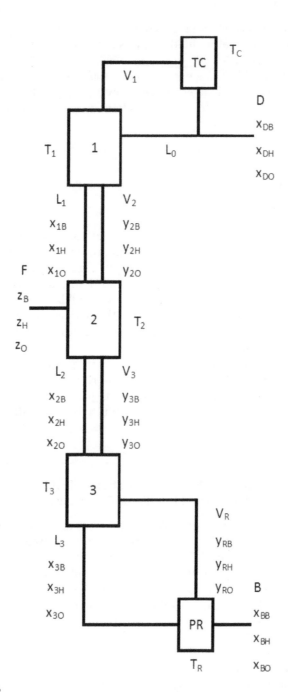

FIGURE 10.2
Three-tray column with TC and PR to separate a three-component mixture of butane, hexane, and octane for use in Example 10.1.

b. Conduct a degree of freedom analysis.

Remember that the compositions of L_0 and V_1 are the same as D and will not count as separate unknowns.

Unknowns (38): T_R, T_1, T_2, T_3, T_C, V_1, L_0, x_{DB}, x_{DH}, x_{DO}, seven streams (L_1, L_2, L_3, V_2, V_3, V_R, and B with their compositions)

Equations (38):

> 1 process specification ($R = L_0/D$)
>
> 1 equilibrium relationship (bubble point for T_C)
>
> 1 total mole balance ($V_1 = L_0 + D$)
>
> 3 process specifications (EMO for V_R, V_3, V_2)

For each tray and reboiler

> 3 mole balances (total, butane, and hexane)
>
> 3 equilibrium relationships (butane, hexane, and octane)
>
> 2 summation equations

It appears there are zero DOF and the equations can be solved.

c. Write the equations and solve.

> Process specification (R): $L_0 = 2D$
>
> Total mole balance: $V_1 = L_0 + D$
>
> Bubble point: $P = K_B x_{DB} + K_T x_{DT} + K_X x_{DX}$ (this determines T_C)
>
> EMO: $V_1 = V_2 = V_3 = V_R$ (three equations, could use liquid flows instead)

For each stage (1, 2, 3, R):

> Total mole balance: $V_{i+1} + L_{i-1} = V_i + L_i$
>
> Benzene mole balance: $y_{i+1B}V_{i+1} + x_{i+1B}L_{i-1} = y_{iB}V_i + x_{iB}L_i$
>
> Toluene mole balance: $y_{i+1T}V_{i+1} + x_{i+1T}L_{i-1} = y_{iT}V_i + x_{iT}L_i$
>
> Benzene equilibrium: $y_{iB} = K_{iB}x_{iB}$ ($K = p^*/P$ assuming an ideal liquid phase)
>
> Toluene equilibrium: $y_{iT} = K_{iT}x_{iT}$
>
> o-xylene equilibrium: $y_{iX} = K_{iX}x_{iX}$

For each of eight streams:

Summations: $\sum x_i = 1$, $\sum y_i = 1$

These are inputted in Excel "BHO/3-Tray." A distillate with 66.7 mol% butane, 31.4 mol% hexane, and 2.0 mol% octane is formed. The bottoms has <0.1 mol% butane, 2.9 mol% toluene, and 97.1 mol% octane. Pencil in all temperatures, flow rates, and compositions in Figure 10.2. Pick representative equations and verify that the values satisfy the equations by hand calculations.

The flow rates, temperature, and compositions at each stage are presented as profiles in Figures 10.3–10.6. The solution by the process simulator ChemCAD using the non-random two-liquid (NRTL) equation for the K values is also plotted for comparison. Notice a maximum for toluene in the middle of the column. Interesting! We'll talk about that later.

d. Explore the problem.

Try some other specifications. For instance, add that 95% of the octane fed is found in the bottoms (instead of 97%). You'll have to vary another parameter to the variables so add R. You'll find that R decreases to 0.29. Now make it 97.5% recovery and you'll find that R increases to 5.5. The percent recovered is very sensitive to the reflux ratio! Now increase the recovery to 99%. Does the simulation (figuratively) "blow up?" Why? Values for process variables do have limits.

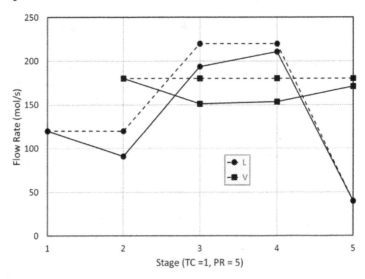

FIGURE 10.3
Flow profiles for the three-tray column in Example 10.1 from Excel simulation (dashed lines) and ChemCAD (solid lines).

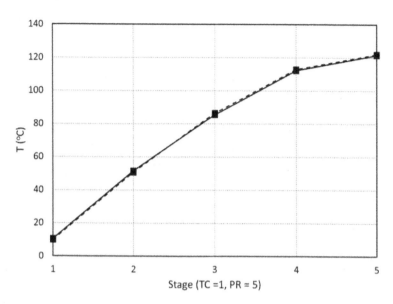

FIGURE 10.4
Temperature profiles for the three-tray column in Example 10.1 from Excel simulation (dashed lines) and ChemCAD (solid lines).

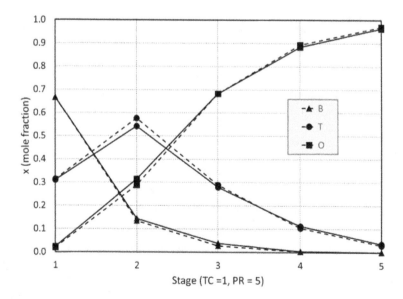

FIGURE 10.5
Liquid composition profiles for the three-tray column in Example 10.1 from Excel simulation (dashed lines) and ChemCAD (solid lines).

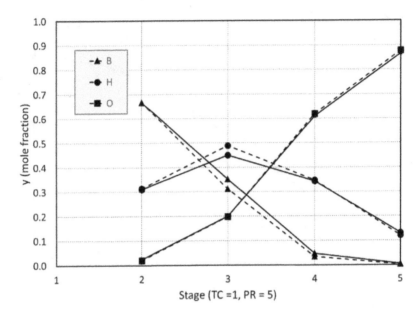

FIGURE 10.6
Vapor composition profiles for the three tray column in Example 10.1 from Excel simulation (dashed lines) and ChemCAD (solid lines).

e. Be awesome!
This challenge is quite awesome indeed! Use Excel "BHOD/DIY" or modify Excel "BHO/3-Tray" to add decane as a fourth component with 100 mol/s of an equimolar feed. The Antoine's constants are included in the data for your convenience. You can compare your results to Excel "BHOD/3-Tray" and accompanying graphs.

**EXAMPLE 10.2   Distillation of BTX in
a Ten-Tray Distillation Column**

A common refinery stream contains benzene, toluene, and xylenes (BTX). The three isomers of xylene are represented here by ortho-xylene. A total of 100 mol/s of a nearly equimolar, saturated liquid mixture of BTX (33% B, 34% T, and 33% X) is fed to the fifth tray of a ten-tray distillation column equipped with a TC and PR. An equal split of distillate and bottoms is produced. The reflux ratio is 2. Assume EMO and perfectly efficient trays.

a. Draw a labeled process flow diagram.

A PFD of the column is shown in Figure 10.7. The feed flow rate and composition are labeled along with the distillate flow rate. D is known from the process description of an equal split.

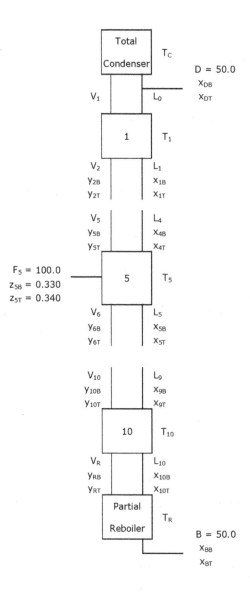

FIGURE 10.7
PFD of a ten-tray distillation column for separating benzene, toluene, and o-xylene for use in Example 10.2.

b. Conduct a degree of freedom analysis.

　　We will guess that $x_{DB} = 0.6$ and $x_{DT} = 0.3$. The o-xylene mole fraction will not be labeled so we don't need the summation equations. Remember that V_1 and L_0 have the same composition as D.

TC/Split Point

Unknowns (3): L_0, V_1, T_C

Equations (3):

　　1 mole balance (total)

　　1 equilibrium relationship (bubble point to calculate T_C)

　　1 process specification (R = 2)

There are zero DOF and this part of the problem can be solved.

Stage 1

Unknowns (7): V_2, y_{2B}, y_{2T}, L_1, x_{1B}, x_{1T}, T_1

Equations (7):

　　3 mole balances (total, B, T)

　　3 equilibrium relationships (B, T, X)

　　1 process specification (EMO)

There appears to be zero DOF and the problem is solvable. Each stage can be analyzed sequentially starting from the top. At the bottom, x_{BB} and x_{BT} from the stage-by-stage calculations and from the overall process mole balances should be the same. If not, new guesses for x_{DB} and x_{DT} are made.

c. Write the equations and solve.

TC/Split Point

Total mole balance: $D + L_0 = V_1$

Bubble point: $K_{DB}x_{DB} + K_{DT}x_{DT} + K_{DX}(1 - x_{DB} - x_{DT}) = 1$

Process specification (reflux ratio): $R = L_0/D$

The results of this calculation are shown below. Specified values are shaded.

Stage	L (mol/s)	x_B (mol B/mol)	x_T (mol T/mol)	V (mol/s)	y_B (mol B/mol)	y_T (mol T/mol)	T (°C)
TC	100	0.600	0.300				90.5
1				150	0.600	0.300	

Since streams are labeled by the stage they are leaving, V_1, y_{1B}, and y_{1T} appear in the row for Stage 1.

Stage 1

Total mole balance: $V_2 + L_0 = V_1 + L_1$

Benzene mole balance: $V_2 y_{2B} + L_0 x_{DB} = V_1 y_{1B} + L_1 x_{1B}$

Toluene mole balance: $V_2 y_{2T} + L_0 x_{DT} = V_1 y_{1T} + L_1 x_{1T}$

Benzene equilibrium: $y_{1B} = K_{1B} x_{1B}$

Toluene equilibrium: $y_{1T} = K_{1T} x_{1T}$

Dew point: $y_{1B}/K_{1B} + y_{1T}/K_{1T} + (1 - y_{1B} - y_{1T})/K_{1X} = 1$

EMO: $V_2 = V_1$

The results of the Stage 1 calculation are shown below. Those values specified and calculated in the previous step are shaded.

Stage	L (mol/s)	x_B (mol B/mol)	x_T (mol T/mol)	V (mol/s)	y_B (mol B/mol)	y_T (mol T/mol)	T (°C)
TC	100	0.600	0.300				90.5
1	100	0.303	0.363	150	0.600	0.300	104.0
2				150	0.402	0.342	

These equations are repeated stage-by-stage until the PR is reached. The flow rates, compositions, and temperature for each stage are shown in Table 10.1. The mole fractions of benzene and toluene from stage-by-stage calculations are compared to those from an overall process mole balance. As expected for this first iteration, they don't match and in fact are negative values. The values for $x_{DB} = 0.6$ and $x_{DT} = 0.3$ were not correct and a new guess is required. We'll let Excel Solver do the work for us by making those differences constraints and making x_{DB} and x_{DT} variables.

The final solution is shown in Table 10.2 and the flow, temperature, liquid mole fraction, and vapor mole fraction profiles are shown in Figures 10.8–10.11. Also shown in the figures are the solutions from the process simulator ChemCAD using the NRTL equation for K values.

TABLE 10.1

Results from First Pass of BTX with Guessed Values of x_B and x_T

Stage	L (mol/s)	x_B (mol B/mol)	x_T (mol T/mol)	V (mol/s)	y_B (mol B/mol)	y_T (mol T/mol)	T (°C)
TC	100	0.600	0.300				90.5
1	100	0.303	0.363	150	0.600	0.300	104.0
2	100	0.147	0.288	150	0.402	0.342	116.8
3	100	0.090	0.199	150	0.298	0.292	124.8
4	100	0.072	0.143	150	0.260	0.232	128.7
5	200	0.066	0.115	150	0.248	0.195	130.4
6	200	0.014	0.012	150	0.068	0.027	141.7
7	200	0.000	−0.044	150	−0.001	−0.111	146.8
8	200	−0.004	−0.070	150	−0.020	−0.185	148.8
9	200	−0.004	−0.082	150	−0.025	−0.220	149.6
10	200	−0.005	v0.087	150	−0.026	−0.235	150.0
PR	50	−0.005	−0.089	150	−0.026	−0.242	150.1
X_R by OMB		0.060	0.360				
Δ =		−0.065	−0.449				

TABLE 10.2

Final Results for BTX

Stage	L (mol/s)	x_B (mol B/mol)	x_T (mol T/mol)	V (mol/s)	y_B (mol B/mol)	y_T (mol T/mol)	T (°C)
TC	100	0.659	0.339				87.8
1	100	0.436	0.556	150	0.659	0.339	94.1
2	100	0.295	0.681	150	0.510	0.484	99.0
3	100	0.220	0.723	150	0.416	0.567	102.3
4	100	0.180	0.696	150	0.366	0.595	105.2
5	200	0.152	0.610	150	0.339	0.577	108.7
6	200	0.083	0.667	150	0.202	0.700	112.3
7	200	0.042	0.689	150	0.110	0.776	114.8
8	200	0.020	0.675	150	0.056	0.806	117.0
9	200	0.009	0.613	150	0.027	0.787	119.6
10	200	0.004	0.490	150	0.012	0.704	123.9
PR	50	0.001	0.321	150	0.004	0.540	130.0
X_R by OMB		0.001	0.321				
Δ =		0.000	0.000				

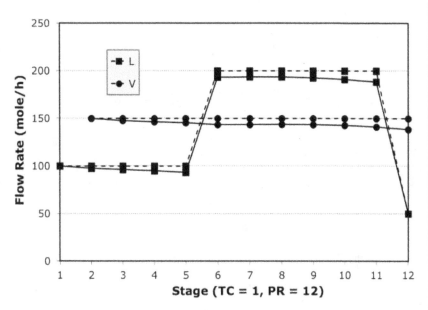

FIGURE 10.8
Flow rate profiles for Example 10.2 with Lewis method results shown with dashed lines and ChemCAD results shown with solid lines.

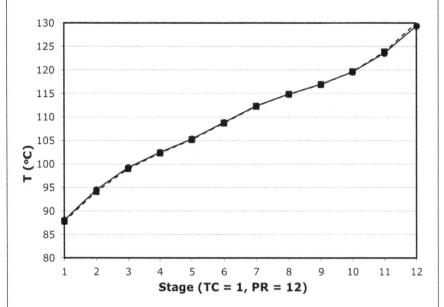

FIGURE 10.9
Temperature profiles for Example 10.2 with Lewis method results shown with dashed lines and ChemCAD results shown with solid lines.

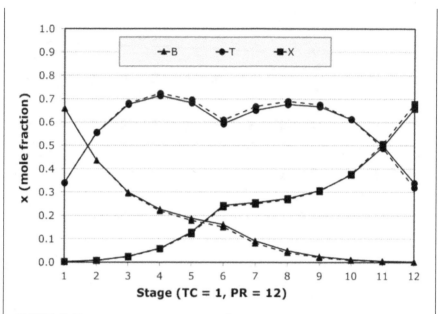

FIGURE 10.10
Liquid composition profiles for Example 10.2 with Lewis method results shown with dashed lines and ChemCAD results shown with solid lines.

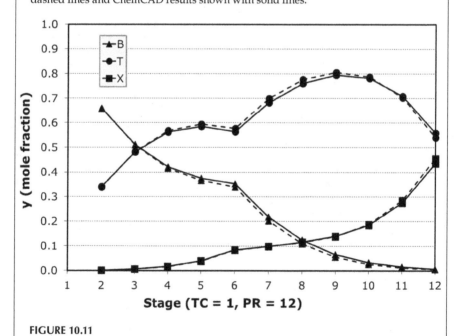

FIGURE 10.11
Vapor composition profiles for Example 10.2 with Lewis method results shown with dashed lines and ChemCAD results shown with solid lines.

d. Explore the problem.

Let's talk about the profiles. The results of this simulation using the Lewis method and the EMO assumption track well with the ChemCAD results. In fact, our experience in Chapter 9 leads us to expect that if we used energy balances instead of the EMO assumption, they would track even better with the ChemCAD results. But there are differences, most easily seen in the flow rate profiles. The ChemCAD generated flow rates are slightly less than found by the Lewis method and decline consistently down the column – except at the feed stage, of course.

The temperature increases smoothly down the column as the flows become richer in the heavier components. A significant kink in the temperature profile would indicate that a non-optimum tray was used for the feed. Excel "BTX/10-Tray" is designed for you to choose different feed trays. Try Trays 4 and 6 to see if the separation has improved at all.

The most curious feature is in the mole fraction profiles, in which the middle boiling toluene composition has an "M" shaped curve with maximums on Trays 3 and 7 for the liquid and Trays 4 and 8 for the vapor. Just for fun, ask your TA for a simple explanation for this shape. Now, let me give it a try.

The benzene and xylene profiles look pretty much like you would expect in two-component distillation – these are the lightest and heaviest components. The benzene mole fraction increases from the bottom to the top of the column. The opposite is the case for xylene. The toluene profile has two maximums and a dip at the feed stage – very strange! What's going on here? The profiles are determined by the primary components being distilled in each section of the column. Let's start with the bottom (Stage 12, the PR) and go up the column.

Stages 12–9 (PR to Tray 8)
Very little benzene is present in the reboiler, so the distillation is mainly between toluene (light) and xylene (heavy). The toluene increases rapidly (going up in the column) and the xylene decreases rapidly. However, the xylene can only decrease so far because a significant amount is added at the feed stage. After its rapid decrease, the composition levels off around Stage 9.

Stages 9–6 (Tray 8 to feed tray)
At Stage 9, significant amounts of benzene start to show up. Since the xylene is not changing (much) the distillation is now mainly between benzene (light) and toluene (heavy). (Notice

how toluene changed from being considered the light to the heavy component?) So the benzene increases as the light component and the toluene decreases as the heavy component. The result is that first maximum on the right hand side of the profile.

Stages 6–4 (feed stage to Tray 3)
All components were added at the feed stage, but now the xylene composition can fall off rapidly to near zero since it is the heaviest component. Both the toluene and benzene are lighter and increase until Stage 4.

Stages 1–4 (TC to Tray 3)
At this point, the xylene is almost all gone and the distillation becomes mainly between the benzene (light) and toluene (heavy, once again!). The benzene continues to increase and the toluene decreases. The result is the second toluene maximum at Stage 4.

Notice that the distillate contains both benzene and toluene while the bottom contains both toluene and xylene. As a very general rule, you will need three columns to separate three components. The first column effectively removes xylene from the distillate and benzene from the bottoms. Now the distillate must be processed in a second column to separate the benzene from the toluene. The bottoms must be processed in a third column to separate the toluene from the xylene. The toluene from both columns can, of course, be combined.

Are we are still on part (d) Explore the problem?! Go ahead and check out Excel "BTX/10-Tray" and change the process specifications using the suggested experiments and anything you find interesting. Be sure you can explain each result.

e. Be awesome!
See if you can reproduce this simulation using Excel "BTX/DIY." Make sure your results match the simulation.

e+. Be even more awesome!
Even though we used the EMO assumption rather than energy balances, you can still calculate the heat removed from the condenser and heat added to the PR. See if you can estimate Q_C and Q_R (kJ/s).

TRY THIS AT HOME 10.1   Four Component Distillation

A total of 100 mol/s of a saturated liquid is fed to a ten-tray distillation column with a PR and a TC. The feed is an equimolar mixture
of benzene, toluene, ethyl benzene, and o-xylene. The reflux ratio
is 2 and the distillate molar flow rate is the same as the bottoms.
Assume perfectly efficient trays and equimolal overflow.

a. Draw a labeled process flow diagram.
   The PFD is shown in Figure 10.12. Fill in all known values
   from the problem description.

b. Conduct a degree of freedom analysis.

   Condenser/split point:
   Start right at the top with the TC/split point. We already
   know that $D = 50$ mol/s from the process description. Can
   you fill in the unknowns and equations?

Unknowns (6):

Equations (3):

It looks like there are three DOF and just like the BTX
system in Example 10.2, we'll have to guess some compositions. For the BTX system we had to guess two mole
fractions. Here we need to guess three! We can start
with most of the benzene ($x_{DB} = 0.6$), some of the toluene
($x_{DT} = 0.3$), and just a little ethyl benzene ($x_{DE} = 0.1$) in the
distillate. The mole fraction of o-xylene is automatically
calculated ($x_{DX} = 0.0$) using the summation equation. Now
there are zero DOF.

Tray 1
Now we can move down to Tray 1. L_0, V_1, and their compositions are now known. Can you fill in these unknowns
and equations?

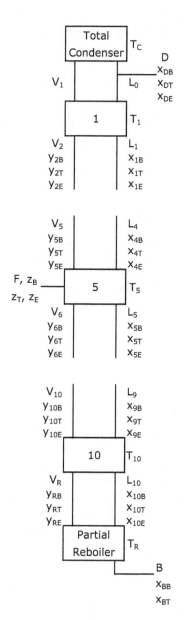

FIGURE 10.12
PFD for ten-tray BTEX distillation column for Try This at Home 10.1.

Unknowns (9):

Equations (9):

There are zero DOF so these equations can be solved. Now the calculations can proceed down the column. At the bottom, the mole fractions for benzene, toluene, and ethyl benzene can be compared to the overall mole balance. If they don't agree, make new guesses for the distillate compositions, rinse, and repeat!

c. Write the equations and solve.

Tray 1

Total mole balance:

Benzene mole balance:

Toluene mole balance:

Ethyl benzene mole balance:

Benzene equilibrium:

Toluene equilibrium:

Ethyl benzene equilibrium:

o-Xylene equilibrium:

EMO:

The spreadsheet simulation for this problem is Excel "BTEX/10-Tray." The flow, temperature, and composition profiles are shown in Figures 10.13–10.16 along with ChemCAD results for comparison.

d. Explore the problem.

The composition profiles of the compounds with inter-mediate volatility – toluene and ethyl benzene – do not have the pronounced double maxima seen with the BTX system. Do your best to explain the shapes to your study partners. Notice that the Lewis method predicts an ethyl benzene mole fraction greater than the o-xylene mole fraction in the bottom product. The more rigorous ChemCAD solution predicts the more logical opposite result. This might be an artifact that we have calculated the o-xylene composition from a summation equation rather than by its individual equilibrium relationship. The error is not large.

e. Be awesome!

Construct the simulation using Excel "BTEX/DIY." Make sure your solution matches the results in Excel "BTEX/10-Tray."

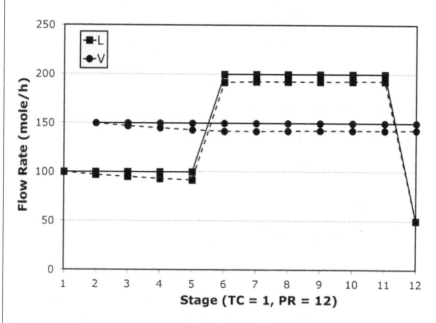

FIGURE 10.13

Flow rate profiles for Try This at Home 10.1 with Lewis method results shown with dashed lines and ChemCAD results shown with solid lines.

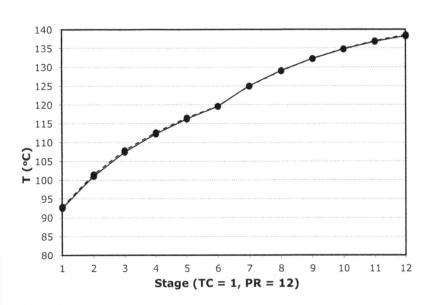

FIGURE 10.14
Temperature profiles for Try This at Home 10.1 with Lewis method results shown with dashed lines and ChemCAD results shown with solid lines.

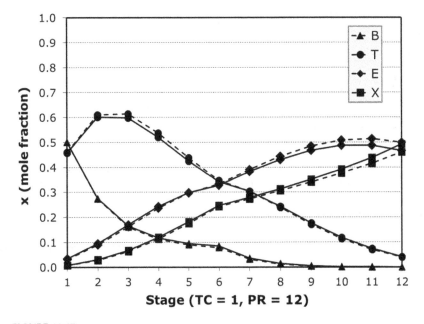

FIGURE 10.15
Liquid composition profiles for Try This at Home 10.1 with Lewis method results shown with dashed lines and ChemCAD results shown with solid lines.

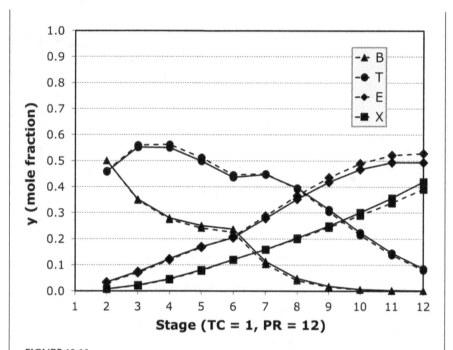

FIGURE 10.16
Vapor composition profiles for Try This at Home 10.1 with Lewis method results shown with dashed lines and ChemCAD results shown with solid lines.

10.3 Summary and Discussion

Multicomponent distillation can be analyzed exactly like binary distillation using mole balances, equilibrium relationships, and various process specifications, most notably, equimolal overflow. This is only practical for relatively simple systems but is clearly useful in this introductory text. Approximate methods such as the Fenske–Underwood–Gilliland method were developed prior to the advent of process simulators (Seader et al., 2015, Chapter 9). Also, a rich history of rigorous methods involving matrix solutions also exists (Seader et al., 2015, Chapter 10). These techniques are best left for advanced study and most chemical engineers will rely on process CAD programs to design complex distillation towers.

11

Distillation of Non-Ideal Systems

Distillation of non-ideal mixtures of chemicals is analyzed just like ideal mixtures except an appropriate activity coefficient model other than Raoult's law is used to describe vapor–liquid equilibrium.

$$\varphi_i y_i P = \gamma_i x_i P_i^* \tag{11.1}$$

φ is the vapor-phase fugacity coefficient, y and x are mole fractions, P is the total pressure, γ is the liquid-phase activity coefficient, and P^* is the vapor pressure. Typically, the vapor phase will be considered ideal so the fugacity coefficient, φ_i, is 1. There are many models for the activity coefficient, γ_i, as discussed in Chapter 2. The one that most closely fits the experimental vapor-liquid equilibrium (VLE) is chosen. Figure 11.1 shows the experimental VLE for ethanol/water mixtures at 1 atm. Also, the VLE calculated with different activity coefficient models – Raoult, Wilson, Margules, and van Laar – are plotted for comparison. This is the same graph as Figure 2.3.

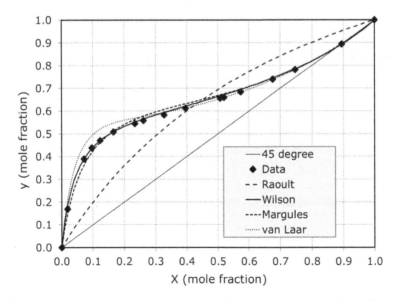

FIGURE 11.1

Vapor–liquid equilibrium data for ethanol–water mixtures at 1 atm (squares) fitted with various activity coefficient models. The Wilson equation best fits the data.

Note that an azeotrope exists near x = y = 0.88 mol E/mol. Raoult's law is a poor fit of the data and cannot model azeotropes. The rest do a good job at the extremes as x_E approaches 0 or 1 and at the azeotrope. The Wilson equation does the best job of modeling ethanol–water VLE at all compositions.

The Wilson equation is

$$\ln \gamma_i = -\ln\left(x_i + x_j\Lambda_{ij}\right) + x_j\left(\frac{\Lambda_{ij}}{x_i + x_j\Lambda_{ij}} - \frac{\Lambda_{ji}}{x_j + x_i\Lambda_{ji}}\right) \qquad (11.2)$$

The Wilson coefficients for ethanol and water are $\Lambda_{EW} = 0.2002$ and $\Lambda_{WE} = 0.8156$. Calculations of VLE with these activity coefficient models are done in Excel "EW/VLE."

11.1 Lewis Method

The analytical Lewis solution is complicated by needing to calculate the activity coefficient using the liquid composition, which might not be known, thus creating a circular reference in Excel. Let's explore this with an example.

EXAMPLE 11.1   Distillation of Ethanol and Water in a 10-Tray Column

A total of 100 mol/s of a saturated liquid containing 40 mol% ethanol and 60 mol% water is fed to Tray 9 of a ten-tray distillation column equipped with a total condenser (TC) and a partial reboiler (PR). About 46 mol/s of distillate is produced with a boilup ratio of 3.0. Assume perfectly efficient stages and equimolal overflow. Determine all flow rates, compositions, and temperatures.

As it will turn out, the very large relative volatility of ethanol at low concentrations makes the optimum feed tray rather low on the column. We'll first analyze the top of the column and find that it is impossible to proceed with the information at hand. Then, we'll start at the bottom of the column and apply the Lewis method from there.

Top of the Column

a. Draw a labeled process flow diagram.
   Both the top and bottom stages are shown in Figure 11.2. Pencil in any values known from the problem statement.
b. Conduct a degree of freedom analysis for the TC/split point.
   The streams crossing this boundary are D, L_0, and V_1.

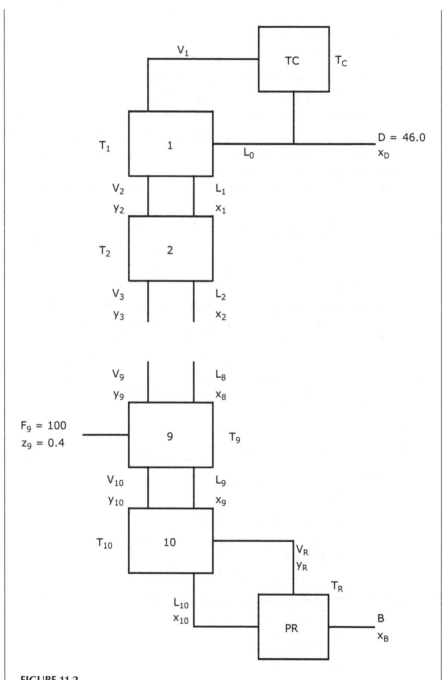

FIGURE 11.2
Process flow diagram for the top and bottom of the ten-tray distillation column in Example 11.1.

Unknowns (4): L_0, V_1, x_D, T_C

Equations (3):

   1 mole balance (total, the others are not independent)

   1 equilibrium relationship (bubble point to find T_C)

   1 process specification (reflux ratio)

There is one DOF. We cannot arbitrarily specify what product composition can be achieved in this specific column. However, as before, we can guess one variable, $x_D = 0.80\,\text{mol/s}$, and proceed. We'll correct that later if the stage-by-stage and overall mole balances don't agree.

c. Write the equations and solve for the TC/split point.

Mole balance: $V_1 = D + L_0$

Equilibrium relationship: $K_E x_D + K_W(1 - x_D) = 1$

Process specification: $L_0 = RD$

Because we have a guessed value for x_D, we are able to calculate the K values at the first guessed temperature. The results are shown below from Excel "EW/10-Tray." So far, so good!

Stage	L (mol/s)	x (mol E/mol)	V (mol/s)	y (mol E/mol)	γ_E	γ_W	K_E	K_W	T (°C)
TC	116	0.800			1.031	2.084	1.024	0.903	78.1
1			162	0.800					

b. Conduct a degree of freedom analysis for Tray 1.

Unknowns (5): V_2, y_2, L_1, x_1, T_1

Equations (5):

   2 mole balances (total, ethanol)

   2 equilibrium relationships (ethanol and dew point or ethanol and water)

   1 process specification (EMO)

With zero DOF we can proceed with the calculations – or can we?

c. Write the equations and solve for Tray 1.

Total mole balance: $V_2 + L_0 = V_1 + L_1$

Ethanol mole balance: $V_2 y_2 + L_0 x_D = V_1 y_1 + L_1 x_1$

Ethanol equilibrium: $y_1 = \left(\gamma_E P_E^*/P\right)x_1$

   γ_E is a function of x_1 and both γ_E and P_E^* are functions of T_1

Dew point: $y_1/(\gamma_E P_E^*/P) + (1 - y_1)/(\gamma_W P_W^*/P) = 1$

EMO: $V_2 = V_1$

I wrote the K values in their expanded form to illustrate a problem. The liquid mole fraction depends on equilibrium and therefore the activity coefficient. But the activity coefficient (from the Wilson equation) depends on the liquid mole fraction. If you try solving these equations in Excel, it will indicate a circular reference. Perhaps if we try starting at the bottom of the column this problem can be avoided.

Bottom of the Column

a. Draw a labeled process flow diagram.

A PFD of the bottom of the column is also drawn in Figure 11.2. An overall process mole balance shows that the bottoms flow rate $B = 54$ mol/s and composition is 0.059 mol B/mol.

b. Conduct a degree of freedom analysis for the reboiler.

Unknowns (6): L_{10}, x_{10}, V_R, y_R, T_R, x_B

Equations (5):

2 mole balances (total, ethanol)

2 equilibrium relationships (ethanol, bubble point)

1 process specification (R_B)

It is no surprise that this approach also requires guessing a bottoms concentration. In Excel "EW/10-Tray," x_D is actually guessed to be 0.80 and x_B is calculated from an overall process mole balance to be 0.059 – effectively the same thing.

c. Write the equations and solve for the reboiler.

Total mole balance: $L_{10} = V_R + B$

Ethanol mole balance: $L_{10}x_{10} = V_R y_R + Bx_B$

Ethanol equilibrium: $y_R = K_{ER}x_B$

Bubble point: $K_{ER}x_B + K_{WR}(1 - x_B) = 1$

Process specification: $V_R/B = R_B$

The activity coefficients can be calculated because we have a value for x_B (albeit guessed). A guess of $x_D = 0.800$ results in $x_B = 0.059$ from an overall process mole balance. In Excel "EW/10-Tray," use the "Goal Seek" function to change T_R until the cell with the bubble point equation is zero. The result is shown below with the specified and guessed values shaded.

Stage	L (mol/s)	x (mol E/mol)	V (mol/s)	y (mol E/mol)	γ_E	γ_W	K_E	K_W	T (°C)
10	216	0.284							
PR	54	0.059	162	0.359	3.978	1.012	6.054	0.682	89.3

b. Conduct a degree of freedom analysis for Tray 10.

Unknowns (5): V_{10}, y_{10}, L_9, x_9, T_{10}

Equations (5):

2 mole balances (total, ethanol)

2 equilibrium relationships (ethanol, bubble point)

1 process specification (EMO)

c. Write the equations and solve for Tray 10.

Total mole balance: $L_9 + V_R = L_{10} + V_{10}$

Ethanol mole balance: $L_9 x_9 + V_R y_R = L_{10} x_{10} + V_{10} y_{10}$

Ethanol equilibrium: $y_{10} = K_{E10} x_{10}$

Bubble point: $K_{E10} x_{10} + K_{W10}(1 - x_{10}) = 1$

EMO: $V_{10} = V_R$

Notice that the liquid-phase mole fraction for a tray will always be known from the calculations for the tray below. That way, the activity coefficients can be calculated and equilibrium equations solved.

In Excel "EW/10-Tray," use the "Goal Seek" function to change T_{10} until the cell with the bubble point equation is zero. The result is shown below with values from the previous calculations or specifications shaded.

Stage	L (mol/s)	x (mol E/mol)	V (mol/s)	y (mol E/mol)	γ_E	γ_W	K_E	K_W	T (°C)
9	216	0.446							
10	216	0.284	162	0.575	1.768	1.179	2.027	0.593	81.8
PR	54	0.059	162	0.359	3.978	1.012	6.054	0.682	89.3

Now the calculations can be made sequentially up the column. At the top of the column, the distillate composition ($y_1 = x_D = x_0$) is compared to the guessed value and a new guess is made if they are not equal. Since Excel calculates from left to right and top to bottom, these calculations are made with a macro (ctrl + shift + B) that performs the bubble point calculations starting at the bottom. As it adjusts the starting guess for x_D, keep running the macro several times until no changes in the bubble point cells are noted. The McCabe–Thiele diagram and flow, temperature, and composition profiles are automatically updated.

The results of the first iteration are shown in Table 11.1 and the final results in Table 11.2. The profiles are shown in Figure 11.3. ChemCAD results are also shown on the profile graphs for comparison.

TABLE 11.1

First Pass of Lewis Method for Example 1.1

Stage	L	x	V	y	γ_E	γ_W	K_B	K_T	T
TC	116	0.866			1.013	2.272	1.003	0.980	78.0
1	116	0.863	162	0.866	1.014	2.262	1.004	0.976	78.0
2	116	0.836	162	0.845	1.021	2.182	1.011	0.943	78.1
3	116	0.809	162	0.826	1.028	2.108	1.021	0.912	78.1
4	116	0.78	162	0.806	1.038	2.033	1.033	0.882	78.2
5	116	0.748	162	0.786	1.051	1.952	1.050	0.851	78.3
6	116	0.709	162	0.763	1.070	1.859	1.077	0.816	78.5
7	116	0.656	162	0.735	1.102	1.744	1.120	0.773	78.7
8	116	0.578	162	0.697	1.166	1.594	1.205	0.719	79.1
9	216	0.446	162	0.641	1.339	1.380	1.437	0.648	80.1
10	216	0.284	162	0.575	1.768	1.179	2.027	0.593	81.8
PR	54	0.059	162	0.359	3.978	1.012	6.054	0.682	89.3

TABLE 11.2

Final Pass of Lewis Method for Example 1.1

Stage	L	x	V	y	γ_E	γ_W	K_B	K_T	T
TC	116	0.824			1.024	2.148	1.015	0.929	78.1
1	116	0.806	162	0.824	1.029	2.100	1.022	0.909	78.1
2	116	0.788	162	0.811	1.035	2.052	1.029	0.890	78.2
3	116	0.768	162	0.798	1.043	2.000	1.039	0.869	78.2
4	116	0.745	162	0.784	1.053	1.943	1.052	0.847	78.3
5	116	0.717	162	0.767	1.066	1.878	1.070	0.822	78.4
6	116	0.682	162	0.747	1.086	1.799	1.096	0.792	78.5
7	116	0.632	162	0.722	1.120	1.695	1.142	0.755	78.8
8	116	0.556	162	0.686	1.189	1.553	1.235	0.704	79.3
9	216	0.422	162	0.632	1.384	1.346	1.498	0.637	80.3
10	216	0.229	162	0.549	2.022	1.126	2.394	0.585	82.6
PR	54	0.039	162	0.293	4.492	1.006	7.410	0.736	91.5

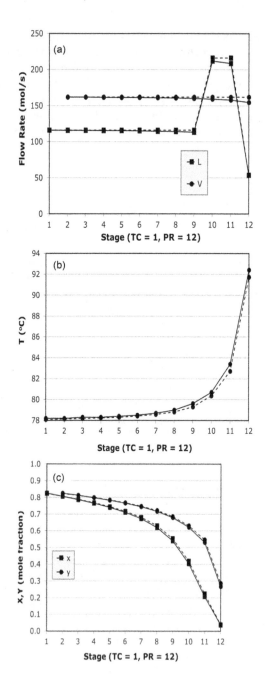

FIGURE 11.3
(a) Flow, (b) temperature, and (c) mole fraction profiles for Example 11.1 with Lewis method results shown with dashed lines and ChemCAD results shown with solid lines.

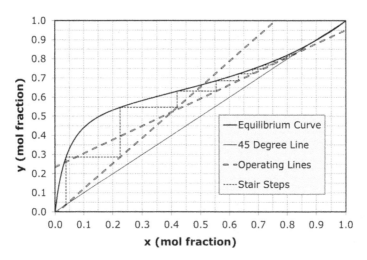

FIGURE 11.4
McCabe–Thiele diagram for Example 11.1. Notice that second tray from bottom is not the optimum feed tray; it should be the last tray.

d. Explore the problem.

On the McCabe–Thiele diagram in Figure 11.4 notice that the passing streams between Trays 10 and 9 are on the BOL but could go a little further on the TOL. That means the optimum feed stage is really Tray 10, the last tray in the column. In Excel "EW/10-Tray" change the 100 mol/s feed and composition 0.4 mol E/mol to Tray 10 and press ctrl+shift+B several times. The stair step between Trays 10 and 9 now "travel" a little further but the distillate purity hasn't changed much because the distillate composition is running up against the azeotrope limit.

You can decrease the ethanol concentration in the bottoms (at the expense of ethanol purity in the distillate) by taking less flow at the bottom. Be careful though – the simulation doesn't like it when your initial guess comes close to pulling ALL of the ethanol out the top (compare Dx_D to Fz). Try $D = 48$ mol/s and an initial guess of $x_D = 0.800$ mol E/mol. It may take several runs of the macro, but you should note a lower x_B. Under these conditions, Tray 9 is the optimal feed tray.

You are in the driver's seat now. Try anything you like, for instance, changing the reflux ratio. If the simulation "blows up," try to figure out why. If the "damage" is not repairable, you can always download the spreadsheet again. For every experiment, note the changes and explain the results to your study partners.

e. Be awesome!

The macro allowed the calculations to proceed from the bottom of the spreadsheet to the top. Macros are programs that you can write in Excel using the Visual Basic programming language. Learn how to do this by watching YouTube "Macros."

e⁺. Be even more awesome!

Thermodynamic data is given in the spreadsheet. Can you include a calculation for the reboiler and condenser heat duties?

11.2 McCabe–Thiele Method

The graphical McCabe–Thiele method can be used exactly as done with ideal systems. The only limitation is that the stair steps cannot go beyond the pinch point formed by an azeotrope.

TRY THIS AT HOME 11.1   McCabe–Thiele Graphical Solution

Repeat Example 11.1 with a graphical McCabe–Thiele solution. Specify that 46 mol/s of distillate is produced with $x_D = 0.823$ and reflux ratio R = 2.52. That is enough information to construct the operating lines and feed line. Print several copies of the xy graph from Excel "EW/xy."

Now you can start at the top and step down, crossing from the TOL to the BOL toward the bottom of the column. However, the operating line and equilibrium line are really close at the top making constructing the stair steps difficult to do accurately. You can avoid this issue by resizing the xy graph as done in Figure 11.5 When the stair steps reach x = 0.6, transition to the normal xy graph and continue. Your result should look like the original graph in Figure 11.4 and reach $x_B = 0.040$ with 11 stages (ten trays and the PR).

Try making the equivalent McCabe–Thiele graph for every scenario you tried in Example 11.1. You can always compare your results with Excel "EW/MT." Are you starting to feel a bit awesome?

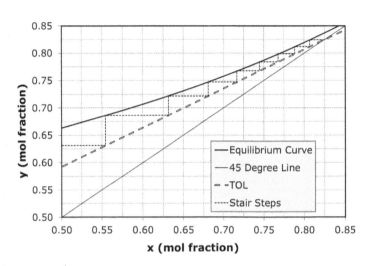

FIGURE 11.5
Zoomed-in portion of the McCabe–Thiele diagram for use in Try This at Home 11.1.

11.3 Summary

Binary systems like ethanol and water are often non-ideal and Raoult's law cannot describe their phase behavior. You can design a column using the graphical McCabe–Thiele analysis exactly like you did for the ideal pair benzene and toluene. The only difference is limited separation because of an azeotrope mixture that forms a pinch point. The analytical Lewis method simply includes calculating an activity coefficient. In the ethanol/water system this is best done with the Wilson equation. The activity coefficients depend on the liquid-phase compositions. The calculation procedure in Excel "EW/10-Tray" requires the calculations be done from the bottom of the column to the top and this is handled here using a macro in the spreadsheet.

Part III

Absorption and Stripping

In Chapters 3–11, we studied distillation, the most common method to separate chemicals. A liquid containing two or more species is partially vaporized by adding heat. The more volatile chemicals concentrate in the vapor while the less volatile chemicals concentrate in the liquid. If a series of equilibrium stages – a distillation column – is used, the separation can be much improved. The important distinction is that the second phase is created from the original phase.

The second most common separation techniques are absorption and stripping. In some ways, these are like distillation in that separation of chemicals occurs between a liquid and a gas phase. However, the second phase is added to – not created from – the original phase. Therefore, adding heat (reboiler) and removing heat (condenser) are usually not necessary.

Absorption is used to transfer chemicals from a gas to a liquid. A liquid is added, intimately contacted with the gas, and then separated. The purpose could be to add a desired chemical to the liquid (for example, carbonating water) or to remove an undesired chemical from the gas (for example, removal of carbon dioxide from natural gas).

Stripping is used to transfer chemicals from a liquid to a gas. A gas is added, intimately contacted with the liquid, and then separated. The purpose could be to add a desired chemical to the gas (for example, humidification of air) or to remove an undesired chemical from the liquid (for example, steam stripping of water to remove volatile organic compounds – VOCs).

If you are thinking that these seem like very similar processes – you are correct. Usually done in a column with trays or packing (like distillation), a liquid is added to the top and physically separated and removed from the bottom. A gas is added to the bottom and physically separated and removed from the top. The difference is in which direction the chemical of interest is transferred – gas to liquid (absorption) or liquid to gas (stripping). Another major difference between absorption/stripping and distillation is the absence of a condenser and reboiler.

We will study these processes in the same way as done for distillation. The process will be built up from a single equilibrium stage in Chapter 12 "Single-Stage Absorption and Stripping," to two stages, and finally to a column with many stages in Chapter 13 "Mathematical Analysis of Absorption/Stripping Columns." This will lead to graphical methods used to design processes in Chapter 14 "Graphical Design of Absorption/Stripping Columns." Two important assumptions in the discussion of absorption and stripping were that only one species changes phase and that the system is isothermal. These are relaxed in Chapter 15 "Multicomponent and Adiabatic Absorption Columns" and it does make a difference. We must continue to distinguish equilibrium stages from actual trays, especially with absorption and stripping that tend to have lower efficiencies than distillation. In Chapter 16 "Column Design," a brief discussion of methods to account for the low efficiencies will be presented along with correlations for determining diameter.

12

Single-Stage Absorption and Stripping

Absorbers and strippers are governed by the same equations: mole balances, equilibrium relationships, and various process specifications. The main difference between the two processes is that the solute moves from liquid to gas in a stripper and the solute moves from gas to liquid in an absorber. In this chapter, we assume that the gas and the liquid leave a single stage at thermodynamic equilibrium. In fact, though, most of these processes are quite inefficient.

12.1 Absorption

The example absorption process will be removal of ethanol from carbon dioxide using water. Ethanol is the solute and water is the absorbent. Large amounts of carbon dioxide are produced in a fermentation process, like brewing beverages or biofuel production. The vented gas is saturated with ethanol (and water, organic sulfides, etc. – but we'll come back to these complications later) and we may want to recover it by absorption into water.

EXAMPLE 12.1   Absorption in a Single Stage

A total of 100.0 mol/s of gas containing 2.00 mol% ethanol (E) and balance carbon dioxide (C) is contacted with pure water (W) in a gas–liquid contactor operating at 30°C and 825 mm Hg. About 95 mol% of the ethanol is absorbed. Determine the flow rates and compositions of all streams.

We can anticipate some complications with this problem and need to make some simplifying assumptions.

1. No specific information is given about this contactor. So we'll assume it acts like an equilibrium stage (meaning the exit streams are in equilibrium) even though most absorption processes are effectively multi-stage and not particularly efficient.

2. Carbon dioxide dissolves in water (think carbonated beverages!) and water evaporates into any dry gas. In this example, assume

---

that only ethanol is transferred between phases. The gas contains only carbon dioxide and ethanol and the liquid contains only water and ethanol. As two-component mixtures, we can simply identify the mole fraction of ethanol in the liquid (x) and in the vapor (y) with the balance being the other component. We will look at multicomponent absorption in Chapter 15.

3. The ethanol–water system is non-ideal and the vapor–liquid equilibrium (VLE) can be described by a modified Raoult's law with the liquid-phase activity coefficient predicted by the Wilson equation. Recall from Chapter 11 that the activity coefficient for infinitely dilute mixtures of ethanol in water is about 6.0. Since this is a dilute system, let's go with that, although it is a very rough estimate as will be seen later.

4. Since the VLE depends on temperature and pressure, we will assume that the system is isothermal and isobaric. Along with a constant activity coefficient (assumed), this makes the K value constant.

   a. Draw a labeled process flow diagram.
      The labeled process flow diagram (PFD) is shown in Figure 12.1.

   b. Conduct a degree of freedom analysis.
      The material balance envelope (MBE) includes all four streams entering and leaving the contactor.

      Unknowns (5): L_{in}, G_{out}, y_{out}, L_{out}, x_{out}

      Equations (5):

         3 mole balances (because there are three species)

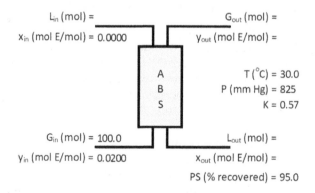

L_{in} (mol) =

x_{in} (mol E/mol) = 0.0000

G_{out} (mol) =

y_{out} (mol E/mol) =

A
B
S

$T\,(^{\circ}C) = 30.0$
P (mm Hg) = 825
$K = 0.57$

G_{in} (mol) = 100.0

y_{in} (mol E/mol) = 0.0200

L_{out} (mol) =

x_{out} (mol E/mol) =

PS (% recovered) = 95.0

FIGURE 12.1
PFD of a single-stage absorber to remove ethanol from carbon dioxide using water as described in Example 12.1.

1 equilibrium relationship (ethanol)

1 process specification (95% recovery)

With zero degrees of freedom (DOF), this problem can be solved.

c. Write the equations and solve.

Ethanol mole balance: $G_{in}y_{in} = L_{out}x_{out} + G_{out}y_{out}$

Carbon dioxide mole balance: $G_{in}(1 - y_{in}) = G_{out}(1 - y_{out})$

Water mole balance: $L_{in} = L_{out}(1 - x_{out})$

Ethanol equilibrium: $y_{out} = Kx_{out}$

Note: $K = y/x = \gamma_E P_E^*/P = (6.00)(78.4)/(825) = 0.570$

Process specification (95%): $0.95 G_{in}y_{in} = L_{out}x_{out}$

We already know x_{in}, G_{in}, y_{in}, T, and P from the problem statement. The equations are solved in Excel "Absorber/1-Stage." The results are L_{in} = 1,061 mol W/s, G_{out} = 98 mol/s, y_{out} = 0.0010 mol E/mol, L_{out} = 1,063 mol/s, and x_{out} = 0.0018 mol E/mol. Go ahead and pencil these values into Figure 12.1 and use them to verify by hand calculations that they satisfy the equations. Note that to make this separation we had no choice in the amount of absorbent. With a series of equilibrium stages, the separation improves so less absorbent is needed or better recovery is achieved.

Since we are not (necessarily) adding or removing heat from the process and are assuming isothermal operation, no energy balance is needed. That's not to say there are never heat considerations in this process. No matter how small the effect, can you think of several that might apply? Make a list and then look at my list in the next example. Heat effects are actually calculated in Chapter 15.

d. Explore the problem.

Use Excel "Absorber/1-Stage" to see what happens when you change the absorbent flow rate or the process specification. This will require small modifications to the spreadsheet, but that should be no problem for you now!

e. Be awesome!

A DIY spreadsheet is included in Excel "Absorber/1-Stage DIY." Input all the equations yourself and make sure you arrive at the same answers for this problem.

TRY THIS AT HOME 12.1   Effect of Temperature and Pressure on Absorption

Ready to give this a try? Use the same problem as Example 12.1 except with only 500 mol water as absorbent. We already observed that there was no choice in the amount of water. Impossible? Not if you allow the temperature to change. Use a lower temperature to increase the equilibrium solubility of ethanol.

  a. Draw a labeled process flow diagram.
     The process flow diagram (PFD) is shown in Figure 12.2. Label what you know and leave the remaining variables blank for now.
  b. Conduct a degree of freedom analysis.
     Let T be one of your unknowns rather than L_{in}.

  Unknowns (5):

  Equations (5):

  You should find five unknowns and five equations for zero DOF and be able to solve the problem.
  c. Write the equations and solve.
     The equations will be the same with the exception that you will need to calculate a new value for K because of its

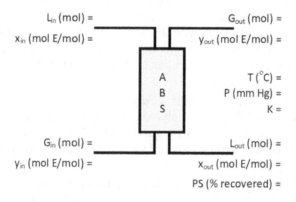

L_{in} (mol) = G_{out} (mol) =
x_{in} (mol E/mol) = y_{out} (mol E/mol) =

                         A T (°C) =
                         B P (mm Hg) =
                         S K =

G_{in} (mol) = L_{out} (mol) =
y_{in} (mol E/mol) = x_{out} (mol E/mol) =
                                 PS (% recovered) =

FIGURE 12.2
PFD of a single-stage absorber to remove ethanol from carbon dioxide using water as described in Try This at Home 12.1.

temperature dependence. See if you can write them down without looking at Example 12.1.

Ethanol mole balance:

Carbon dioxide mole balance:

Water mole balance:

Ethanol equilibrium:

Process specification (95%):

You can solve these equations using Excel "Absorber/1-Stage" with a small modification of the spreadsheet. You will find that the temperature must be 17.2°C. Fill in all the unknown variables in Figure 12.2.

You can use less absorbent with lower temperatures. What are some of the design tradeoffs? Would it matter where in the world or what time of year the process operates?

d. Explore the problem.

Try the same problem but let pressure change rather than temperature. Compare the effect of temperature to pressure and explain any observations. What are some of the design tradeoffs?

e. Be awesome!

You are probably wondering what absorption and stripping equipment looks like. Satisfy that curiosity right now and do an Internet search (be careful with your search terms!). Add schematics and photos to your digital image collection.

12.2 Stripping

Stripping is analyzed much like absorption but the solute will initially be in the liquid and removed by contacting with a gas. The example stripping process will be removing ethanol from water using air. Removing volatile organic compounds from water is a common environmental process but leaves us with the problem of now contaminated air! In all stripping examples, I will demonstrate using the Wilson activity coefficient for non-ideal solutions to check if the infinite dilution assumption was valid.

EXAMPLE 12.2 Single-Stage Stripping

A total of 100.0 mol/s of liquid containing 2.00 mol% ethanol (E) and balance water (W) is mixed with pure air (A) in a gas–liquid contactor operating at 30°C and 825 mm Hg. About 95 mol% of the ethanol is stripped. Determine the flow rates and compositions of all streams.

In addition to the assumptions stated in Example 12.1, note the following:

1. Under these conditions, the gas phase air should behave the same as carbon dioxide. Of course, carbon dioxide will dissolve in water better than air if we were accounting for that.

2. The Wilson activity coefficient (review Chapter 11) is a function of concentration and really should be calculated to determine K values. The equation for γ_E is

$$\gamma_E = \exp\left(-\ln\left(x_E + \Delta_E x_W\right) + x_W \varphi\right) \qquad (12.1)$$

where

$$\varphi = \frac{\Delta_E}{x_E + x_W \Delta_E} - \frac{\Delta_W}{x_W + x_E \Delta_W} \qquad (12.2)$$

$\Delta_E = 0.2002$ and $\Delta_W = 0.8156$ are constants for ethanol and water in the Wilson equation. x_E and x_W are the mole fractions for ethanol and water in the liquid phase. Notice that the Wilson activity coefficient is not a function of temperature, pressure, or vapor-phase mole fraction.

a. Draw a labeled process flow diagram.
   The PFD is shown in Figure 12.3.

b. Conduct a degree of freedom analysis.

   Unknowns (5): G_{in}, G_{out}, y_{out}, L_{out}, x_{out}

   Equations (5):

   > 3 mole balances (because there are three species, as before)
   > 1 equilibrium relationship (ethanol)
   > 1 process specification (95% recovery)

   With zero DOF, this problem can be solved.

c. Write the equations and solve.

   Ethanol mole balance: $L_{in}x_{in} = L_{out}x_{out} + G_{out}y_{out}$

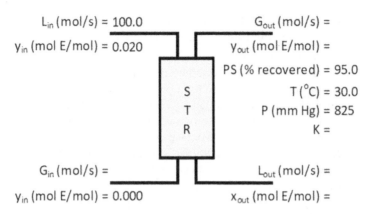

FIGURE 12.3
PFD of single-stage stripper to remove ethanol from water using air as described in Example 12.2.

Air mole balance: $G_{in} = G_{out}(1 - y_{out})$

Water mole balance: $L_{in}(1 - x_{in}) = L_{out}(1 - x_{out})$

Ethanol equilibrium: $y_{out} = Kx_{out}$

Note: $K = y/x = \gamma_E P_E^*/P = (?)(78.4)/(825)$

where γ_E will depend on x_{out}.

Process specification (95%): $0.95L_{in}x_{in} = G_{out}y_{out}$

The equations are solved in Excel "Stripper/1-Stage." The results are $G_{in} = 3{,}290$ mol/s, $G_{out} = 3{,}292$ mol/s, $y_{out} = 0.0006$ mol E/mol, $L_{out} = 98.1$ mol/s, and $x_{out} = 0.0010$ mol E/mol. Go ahead and pencil these values into Figure 12.3 and use hand calculations to verify that they satisfy the equations. Note that in order to make this separation we had no choice in the amount of gas. With a series of equilibrium stages, the process becomes more efficient and less gas is needed or better recovery is achieved.

In part (d) of Example 12.1, I promised to list a few heat considerations. A key assumption in this analysis is that the system is isothermal. You would need to consider if (1) the liquid or gas feed is not at the specified process temperature, necessitating heat exchangers or specific heat calculations, (2) heat of mixing is significant, (3) heat of vaporization is significant, (4) there is heat transfer to the surroundings. Of course, you could drop the isothermal assumption but that complicates the analysis.

d. Explore the problem.

The K value is a function of temperature through the vapor pressure term, the total pressure, and the composition through the activity coefficient. Did you notice that the Wilson activity coefficient at 2.0 mol% is 5.1 – not 6.0, so infinite dilution in Example 12.1 was not a particularly good assumption? Try changing the temperature and pressure to see how that affects the amount of gas needed. Explain your results.

e. Be awesome!

A DIY spreadsheet is included in Excel "Absorber/1-Stage DIY." Input all the equations yourself and make sure you arrive at the same answers for this problem.

TRY THIS AT HOME 12.2   Effect of Temperature and Pressure on Stripping

In Try This at Home 12.1, we saw that the intensive variables, T and P, could affect the operation of an absorber so it makes sense that the same is true for a stripper. Use the same problem as Example 12.2 except with 500 mol air. Use a higher temperature to decrease the equilibrium solubility of ethanol in water so that it comes out easier.

a. Draw a labeled process flow diagram.

The process flow diagram is shown in Figure 12.4. Label what you know from the process description and leave the remaining variables blank.

b. Conduct a degree of freedom analysis.

Let T be one of your unknowns rather than G_{in}.

Unknowns (5):

Equations (5):

You should find zero DOF and be able to solve the problem.

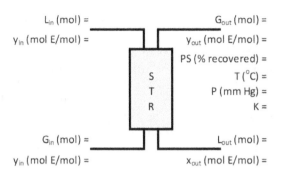

FIGURE 12.4

PFD of a single-stage stripper to remove ethanol from water using air as described in Try This at Home 12.2.

c. Write the equations and solve.

The equations will be the same with the exception that you will need to calculate a new value for K. See if you can write them down without looking at Example 12.1.

Ethanol mole balance:

Carbon dioxide mole balance:

Water mole balance:

Ethanol equilibrium:

Process specification (95%):

You can solve these equations using Excel "Stripper/1-Stage" with a small modification of the spreadsheet. You will find that the temperature must be 68.7°C. Fill in all the unknown variables in Figure 12.4.

You can use less gas with higher temperatures. What are some of the design tradeoffs? Would it matter where in the world or what time of year the process operates?

d. Explore the problem.

Try the same problem but let pressure change rather than temperature. Compare the effect of temperature to the effect of pressure and explain any observations. What are some of the design tradeoffs?

e. Be awesome!

You are pretty close to achieving a state of awesomeness. Reward yourself with a trip to the gym!

12.3 Summary

A chemical is transferred from a gas to a liquid in absorption and from a liquid to a gas in stripping. The major difference with distillation is that the second phase is added to the feed rather than created from the feed. Both processes are governed by mole balances, phase equilibrium relationships, and various process specifications. The example processes used water and ethanol that form a non-ideal solution, well described by Wilson activity coefficients.

13

Mathematical Analysis of Absorption/Stripping Columns

Just like distillation, using multiple stages with countercurrent flow greatly increases the degree of separation in an absorption or stripping process. The types of equations describing multi-stage absorbers and strippers are the same as single-stage processes, only their number increases. The equations are mole balances, phase equilibrium, energy balances (if required), and various process specifications. In this chapter, we progress from two stages to a full seven-stage column. Solute-free compositions and flow rates are introduced to simplify the analysis.

13.1 Two Stages

Let's start with a two-stage absorber using the same specifications as Example 12.1. We should see an improvement in performance, either using less absorbent for the same separation or better separation for the same absorbent flow rate.

EXAMPLE 13.1   Two-Stage Absorber

A total of 100.0 mol/s of gas containing 2.0 mol% ethanol (E) and balance carbon dioxide (C) is contacted with pure water (W) in a two-stage absorber operating at 30°C and 825 mm Hg. About 95 mol% of the ethanol is absorbed. Determine the flow rates and compositions of all streams. Assume that only the ethanol is transferred between phases, the process is isobaric and isothermal, and the activity coefficient is 6.0 ($K = 0.57$ under these conditions).

a. Draw a labeled process flow diagram.

The labeled process flow diagram (PFD) is shown in Figure 13.1. The streams are labeled according to the stage they exit. In the case of gas and liquid feeds, they are labeled as if they came from Stages 3 and 0, respectively.

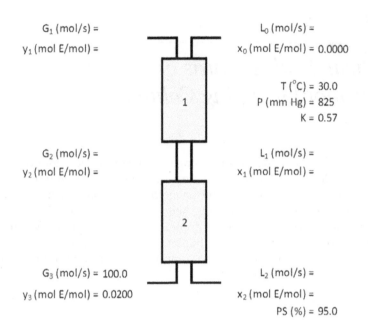

FIGURE 13.1
Process flow diagram of a two-stage absorber as described in Example 13.1.

b. Conduct a degree of freedom analysis.
   Let's use the brute force method in which we solve for all streams at once.

   Unknowns (9): G_1, y_1, G_2, y_2, L_0, L_1, x_1, L_2, x_2

   Equations (9):

   1 process specification (95%)
   For each stage
   3 mole balances (ethanol, water, carbon dioxide)
   1 equilibrium relationship (ethanol, it is the only species distributed)

   There are zero DOF. The problem can be solved.

c. Write the equations and solve.

   Stage 1

   Ethanol mole balance: $G_2 y_2 = G_1 y_1 + L_1 x_1$

   Carbon dioxide mole balance: $G_2(1 - y_2) = G_1(1 - y_1)$

Water mole balance: $L_0 = L_1(1 - x_1)$

Ethanol equilibrium: $y_1 = Kx_1$

Stage 2

Ethanol mole balance: $G_3y_3 + L_1x_1 = G_2y_2 + L_2x_2$

Carbon dioxide mole balance: $G_3(1 - y_3) = G_2(1 - y_2)$

Water mole balance: $L_1(1 - x_1) = L_2(1 - x_2)$

Ethanol equilibrium: $y_2 = Kx_2$ (Note: Continue to assume $K = 0.57$)

Process specification (95%): $0.95G_3y_3 = L_2x_2$

Solve the equations using Excel "Absorber/2-Stage." Note that a single stage required 1,061 mol/s of the absorbent water, while two stages require just 217 mol/s of water. Fill in all values for the variables in Figure 13.1. Use these values in hand calculations to verify the equations are satisfied.

d. Explore the problem.

Use Excel "Absorber/2-Stage" to explore the problem. What happens if you change the percent recovery, amount of absorbent, or process specifications like T or P? Be sure to explain your results to study partners.

e. Be awesome!

Input all equations into Excel "Absorber/2-Stage DIY" and verify that your answers are the same.

TRY THIS AT HOME 13.1   **Two-Stage Stripper**

A total of 100.0 mol/s of liquid containing 2.0 mol% ethanol (E) and balance water (W) is contacted with pure air (A) in a two-stage stripper operating at 30°C and 825 mm Hg. About 95 mol% of the ethanol is stripped. Determine flow rates and compositions of all streams. Assume that only the ethanol is transferred between phases, and the process is isobaric and isothermal.

These are the same specifications used in Example 12.2 involving stripping in a single stage so that we can compare results. Again, we will account for non-ideality in the liquid phase by using Wilson activity coefficients in a modified Raoult's law.

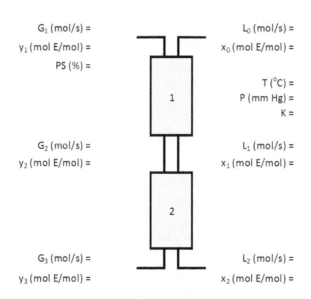

Stage labels:

G_1 (mol/s) =
y_1 (mol E/mol) =
PS (%) =

L_0 (mol/s) =
x_0 (mol E/mol) =

T (°C) =
P (mm Hg) =
K =

G_2 (mol/s) =
y_2 (mol E/mol) =

L_1 (mol/s) =
x_1 (mol E/mol) =

G_3 (mol/s) =
y_3 (mol E/mol) =

L_2 (mol/s) =
x_2 (mol E/mol) =

FIGURE 13.2
Process flow diagram of a two-stage stripper as described in Try This at Home 13.1.

 a. Draw a labeled process flow diagram.
    The PFD is shown in Figure 13.2. Add specified values and leave the remaining variables blank.

 b. Conduct a degree of freedom analysis.
    Use the "brute force" method of solving for all unknowns at once.

    Unknowns (9):

    Equations (9):

    The equilibrium constant K depends on already-identified variables x, T, and P, so it does not count as an unknown. There are zero DOF so the problem can be solved.

 c. Write the equations and solve.

    Stage 1
    Ethanol mole balance:

Carbon dioxide mole balance:

Water mole balance:

Ethanol equilibrium:

Process specification (95%):

Stage 2
Ethanol mole balance:

Carbon dioxide mole balance:

Water mole balance:

Ethanol equilibrium:

Solve these equations in Excel "Stripper/2-Stage DIY." The Wilson activity coefficient equations are included for your convenience. Remember that only the ethanol is in equilibrium – we assumed the water and air did not transfer between phases.

Compare the 686 mol/s of air required here with the 3,290 mol/s of air required for the single stage in Example 12.2. Check your solution with the completed simulation in Excel "Stripper/2-Stage." Fill in all values for unknown variables identified in Figure 13.2. Use hand calculations to verify that the equations are satisfied.

d. Explore the problem.

Use Excel "Stripper/2-Stage" to observe the effect of changing compositions, flow rates, and process specifications on the performance of the stripper. Be sure to discuss the results with your study partners.

e. Be awesome!

The ethanol is changing phase from a liquid to a gas, while, realistically, some water is evaporating into the air. (The amount of air dissolving into the liquid is negligible.) How would this affect the temperature at each stage in a perfectly insulated column? You would be really awesome to do some quick order-of-magnitude calculations. What would it take to maintain an isothermal column? Discuss with your study partners and in class.

13.2 Solute-Free Variables

Before continuing with multi-stage absorption and stripping, it will be convenient to define a solute-free mole fraction. It is the number of moles of the
solute divided by the number of moles of everything else. If each phase is
a simple two-chemical mixture, it amounts to a mole ratio as opposed to a
mole fraction. It will be designated with capitalized letters, and in terms of
the example absorber: X (mol E/mol W) and Y (mol E/mol C). Also, we define
solute-free molar flow rates L′ (mol W/s) and G′ (mol C/s) that are designated
with a prime symbol.

$$G' \equiv (1-y)G \qquad G \equiv \frac{G'}{(1-y)} \tag{13.1}$$

$$Y \equiv \frac{y}{(1-y)} \qquad y \equiv \frac{Y}{(1+Y)} \tag{13.2}$$

$$L' \equiv (1-x)L \qquad L \equiv \frac{L'}{(1-x)} \tag{13.3}$$

$$X \equiv \frac{x}{(1-x)} \qquad x \equiv \frac{X}{(1+X)} \tag{13.4}$$

Why do this? If we assume no water evaporates into the gas and no carbon
dioxide dissolves in the water, solute-free variables simplify the analysis.
Then as the carbon dioxide flows up the column, G′ (mol C/s), and water
flows down the column, L′ (mol W/s), their molar flow rates don't change.
A complete description of the flow and concentration is provided by these
variables. You should try proving these identities to yourself.

For material balance calculations, the amount of ethanol in each stream is
simply

$$n_{gas}(mol\ E/s) = \frac{G'(\cancel{mol\ C})}{s} \left| \frac{Y(mol\ E)}{(\cancel{mol\ C})} \right.$$

$$n_{liq}(mol\ E/s) = \frac{L'(\cancel{mol\ W})}{s} \left| \frac{X(mol\ E)}{(\cancel{mol\ W})} \right.$$

Let's practice using these variables for the two-stage absorber in Example 13.1.

EXAMPLE 13.2 Two-Stage Absorber with Solute-Free Variables

About 95% of the ethanol is to be removed from 100.0 mol/s of a 2.00 mol% mixture of ethanol (E) in carbon dioxide (C) using water (W) in a two-stage absorber operating at 30°C and 825 mm Hg. Assume that only the ethanol is transferred between phases, the process is isobaric and isothermal, and the activity coefficient is 6.0 (K = 0.57 under these conditions). We found the flow rate and composition of every stream in Example 13.1. We will try this again using the solute-free variables.

First, we convert standard to solute-free molar flow rates and mole fractions.

$$G' = (1-y)G = (1-0.0200)(100.0) = 98.0 \text{ mol C/s}$$

$$Y_{in} = \frac{y}{1-y} = \frac{0.0200}{(1-0.0200)} = 0.0204 \text{ mol E/mol C}$$

At these low concentrations, the difference between Y and y is small. In the reverse direction

$$y = \frac{Y}{1+Y} = \frac{0.0204}{1.0204} = 0.0200$$

$$G = \frac{G'}{(1-y)} = \frac{98.0}{(1-0.0200)} = 100.0 \text{ mol/s}$$

a. Draw a labeled process flow diagram.

   The labeled PFD using these solute-free flow rates and concentrations is shown in Figure 13.3. L' and G' are not repeated for each stream because they are assumed constant.

b. Conduct a degree of freedom analysis.

   Use the "brute force" method of solving for all unknowns at once.

   Unknowns (5): L', Y_1, Y_2, X_1, X_2

   Equations (5):

       1 process specification (95% recovery)

           For each stage:

       1 mole balance (ethanol)

       1 equilibrium relationship (ethanol)

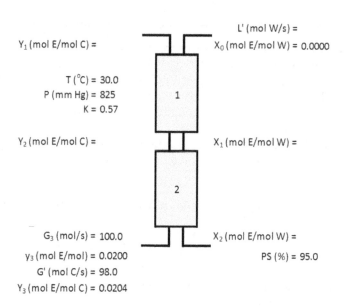

FIGURE 13.3
Process flow diagram of a two-stage absorber with variables expressed in solute-free coordinates as described in Example 13.2.

There are five unknowns and five equations leaving zero DOF and the problem can be solved. The water and carbon dioxide flow rates are assumed constant ($L'_{in} = L'_{out} \equiv L'$, $G'_{in} = G'_{out} \equiv G'$). This eliminates the need for water or carbon dioxide mole balances.

c. Write the equations and solve.

Stage 1

Ethanol mole balance: $G'Y_2 = L'X_1 + G'Y_1$

Ethanol equilibrium: $Y_1/(1 + Y_1) = KX_1/(1 + X_1)$

Note: equilibrium was expressed in mole fractions ($y = Kx$).

Stage 2

Ethanol mole balance: $G'Y_3 + L'X_1 = L'X_2 + G'Y_2$

Ethanol equilibrium: $Y_2/(1 + Y_2) = KX_2/(1 + X_2)$

Process specification (95%): $(0.95)G'Y_3 = L'X_2$

We already know G', Y_3, X_0, and K. The equations are solved in Excel "Absorber/2-Stage SF." Of course, the answers are equivalent to Example 13.1. Prove that to yourself.

d. Explore the problem.

This is like the suggestion in Example 13.1. Use Excel "Absorber/2-Stage SF" to see what happens when you change the absorbent flow rate, temperature, pressure, or the process specification. This will require small modifications to the spreadsheet.

e. Be awesome!

Use the DIY template in Excel "Absorber/2-Stage DIY" to solve this problem by inserting the appropriate equations. Make sure you get the same answers.

13.3 Multi-Stage Absorption Columns

If two stages are better than one stage, then it makes sense that adding even more will result in better separation. Of course, there are always tradeoffs, most notably the larger capital cost of a taller column with more stages versus the lower operating cost using reduced absorbent. (Lower flow rates can also reduce capital cost by allowing smaller diameter columns.)

The "brute force" method will not be practical with this many unknowns so we will use the Lewis method. As with distillation columns, there will typically not be enough information to determine all flow rates and mole fractions for any one stage. So we will make a guess for absorbent flow rate and then proceed with calculations down the column stage by stage. At the end, we will check if the feed gas composition is correct and adjust the absorbent flow rate if needed. This is quite similar to the Lewis method for distillation columns and will be demonstrated in the following example.

EXAMPLE 13.3   Seven-Stage Absorption Column

A total of 100 mol/s of gas containing 2.0 mol% ethanol (E) and balance carbon dioxide (C) is mixed with pure water (W) in a seven-stage absorber operating at 30°C and 825 mm Hg. About 95 mol% of the ethanol is absorbed. Determine the flow rates and compositions of all streams using solute-free mole fractions and flow rates. Assume that only the ethanol is transferred between phases, the process is isobaric and isothermal, and the activity coefficient is 6.0 (K = 0.57 under these conditions). Note that these are the same specifications for the one-stage absorber in Example 12.1 and two-stage absorber in Example 13.1 so that we can compare performances.

a. Draw a labeled process flow diagram.

The PFD is shown in Figure 13.4. Specified values are shown and the remaining solute-free mole fractions and flow rates are indicated. Recall from Example 13.2 that $G' = 98$ mol C and $Y_8 = 0.0204$ mol E/mol C. Pencil in these values.

b. Conduct a degree of freedom analysis.

For this problem, we will use the stage-by-stage Lewis method.

Stage 1

Unknowns (4): L', Y_1, Y_2, X_1 (note that L' and G' are not indicated for every stage)

Equations (3):

    1 mole balance (ethanol)

    1 equilibrium relationship (ethanol)

    1 process specification (5%)

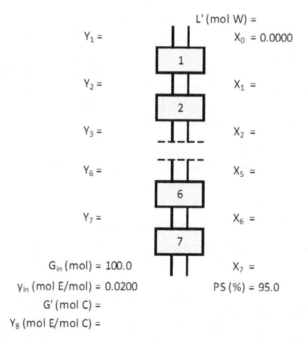

FIGURE 13.4

Process flow diagram of a seven-stage absorber used in Example 13.3.

This leaves one DOF and the problem cannot be solved. Note that the process specification can be applied at either the top or bottom of the column – 95% of the ethanol fed is in the water leaving the bottom or 5% of the ethanol remains in the gas leaving the top. Your choice!

Stage 7

Unknowns (4): L', Y_7, X_6, X_7

Equations (3):

    1 mole balance (ethanol)

    1 equilibrium relationship (ethanol)

    1 process specification (95%)

This also leaves one DOF and the problem cannot be solved. Good luck finding any mole balance envelope that can be solved!

If L' is guessed, Stage 1 (or Stage 7, for that matter) will have zero DOF and the problem can be solved. Then Stage 2 will have zero DOF and can be solved and so on down the column. At the bottom stage, the calculated gas phase composition can be compared to the specified input composition. If they are not the same (and it would be a very lucky initial guess if they are) the absorbent flow rate is changed until they converge.

c. Write the equations and solve.

Stage 1

Guess $L' = G' = 98.0\,mol$

Ethanol mole balance: $G'Y_2 = G'Y_1 + L'X_1$ (Note: G' specified, L' guessed)

Ethanol equilibrium: $Y_1/(1 + Y_1) = KX_1/(1 + X_1)$

Process specification (5%): $G'Y_1 = (0.05)G'Y_8$ (Note: 5 mol% remains in gas)

We already know or guessed L', X_0, Y_8, and Y_1 (from the PS). The ethanol mole balance and equilibrium equations can be solved iteratively but are easily written explicitly for X_1 and Y_2 with a little bit of algebra. For any stage

Ethanol equilibrium: $X_i = F_i/(1 - F_i)$ (where $F_i = Y_i/(K(1 + Y_i))$)

Ethanol mole balance: $Y_{i+1} = Y_i + (L_i/G_i)(X_{i-1} - X_i)$

These equations are solved for each stage and the solution is

Stage	X (mol E/mol W)	Y (mol E/mol C)
(0)	0.0000	
1	0.0018	0.0010
2	0.0049	0.0028
3	0.0105	0.0060
4	0.0204	0.0115
5	0.0382	0.0214
6	0.0709	0.0392
7	0.1335	0.0719
(8)		0.1345
	Y_8 Specified =	0.0204
	Δ	0.1141

Y_8 is the gas feed composition resulting from the stage-by-stage calculation. Unfortunately, it is quite different than the specified composition of 0.02041 mol E/mol C, so we made a really lousy guess for L′. Excel "Absorber/7-Stage" adds the difference (Δ) between Y_8 and the specified composition as a constraint and lets L′ be a variable in Solver. The final solution with L′ = 69.5 mol water is shown below.

Stage	X (mol E/mol W)	Y (mol E/mol C)
(0)	0.0000	
1	0.0018	0.0010
2	0.0040	0.0023
3	0.0068	0.0039
4	0.0103	0.0059
5	0.0147	0.0083
6	0.0203	0.0115
7	0.0273	0.0154
(8)		0.0204
	Y_8 Specified =	0.0204
	Δ	0.0000

1,061, 217, and now 70 mol/s of water is required in the one-, two-, and seven-stage absorbers to remove 95% of the ethanol from the gas. As expected, more stages require less absorbent.

d. Explore the problem.

Check out the effects of T, P, and L′ on the performance of the column using Excel "Absorber/7-Stage." Systematically vary the parameters, report the results to your study group, and discuss.

e. Be awesome!

Try reproducing this simulation using Excel "Absorber/7-Stage DIY." Make sure your results are identical. Even more awesome would be to use Wilson's equation for the ethanol activity coefficient, like we did for the stripper. Does it make a significant difference?

13.4 An Interesting Observation (Again!)

The compositions of the leaving streams (X_i, Y_i) and the passing streams (X_i, Y_{i+1}) are plotted on the XY graph in Figure 13.5. This plot is also found in Excel "Absorber/7-Stage XY" and updated with every trial. A straight line is drawn between points in each set. The passing streams are related by mole

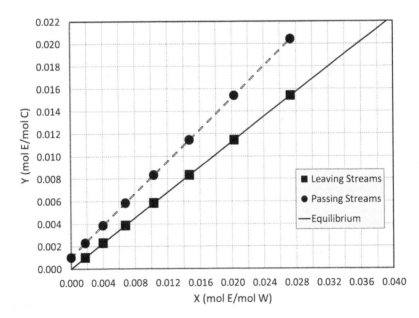

FIGURE 13.5

Passing (circles) and leaving streams (squares) for the seven-stage absorber used in Example 13.3.

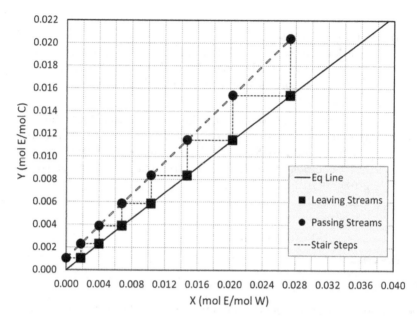

FIGURE 13.6
Demonstration of stepping off stages for the seven-stage absorber used in Example 13.3.

balances resulting in a straight line if the solute-free molar flow rates are indeed constant. As we named it for distillation, this is called the operating line. Notice that for absorption, the operating line lies above the equilibrium line. The leaving streams all fall on the equilibrium curve. It is not necessarily straight and will have a noticeable curve at higher concentrations.

We can draw a line between alternating passing and equilibrium streams. In Figure 13.6, the passing streams at the top of the column (X_0, Y_1) are connected to the equilibrium streams leaving Stage 1 (X_1, Y_1). This in turn is connected to the streams passing between Stages 1 and 2 (X_1, Y_2) and continued to the bottom of the column at (X_7, Y_8).

The result looks like a set of stairs and is completely analogous to the McCabe–Thiele method for distillation column design developed in Chapter 7. Once again, you have discovered a simple, graphical technique to design absorption and stripping columns that will be developed in the next chapter.

13.5 Multi-Stage Stripping Columns

The analysis of multi-stage stripping columns is very similar to absorption columns. The main difference, of course, is that the solute transfers from the liquid to the gas. As you will see, the passing stream compositions will lay on an operating line below the equilibrium line.

EXAMPLE 13.4 Seven-Stage Stripping Column

A total of 100.0 mol/s of liquid containing 2.0 mol% ethanol (E) and balance water (W) is contacted with pure air (A) in a seven-stage stripping column operating at 30°C and 825 mm Hg. About 95 mol% of the ethanol is stripped out of the liquid. Determine the flow rates and compositions of all streams. Assume an isobaric and isothermal process with perfectly efficient stages. These are the same specifications as Example 12.2 and Try This at Home 13.1 for comparison with one- and two-stage processes.

a. Draw a labeled process flow diagram.
   A labeled PFD is shown in Figure 13.7. Add any specified values. Let's go ahead and calculate L' and X_0.

$$L' = L(1 - x_0) = 100(1 - 0.02) = 98.0$$

$$X_0 = \frac{x}{(1-x)} = \frac{0.02}{(1-0.02)} = 0.0204$$

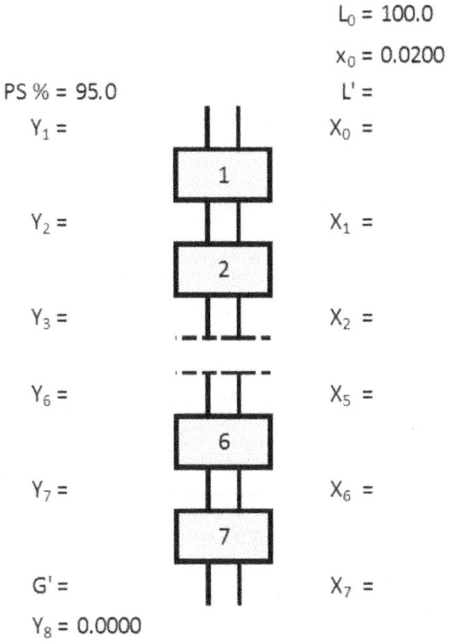

FIGURE 13.7
Process flow diagram for the seven-stage stripper used in Example 13.4.

b. Conduct a degree of freedom analysis.

Stage 1

Unknowns (4): G', Y_1, Y_2, X_1

Equations (3):

    1 mole balance (ethanol)

    1 equilibrium relationship (ethanol)

    1 process specification (95%)

Don't forget the gas molar flow rate that was only labeled at the bottom for convenience. The problem cannot be solved yet with one DOF.

    You know what to do now. We make a reasonable guess for G' and solve the equations for each stage, starting at the bottom and working up the column. Reaching X_0 at the top of the column, we compare the value to that originally specified in the feed liquid. If not the same, make a new guess for G', rinse, and repeat.

    Why start at the bottom? For Stage 7, we can calculate X_7 from an overall process mole balance. Then we can calculate the activity coefficient and determine Y_7 by equilibrium. If instead we tried starting at the top of the column, we wouldn't have X_1 and could not calculate the activity coefficient.

c. Write the equations and solve.

Stage 7
Guess G' to be 100 mole air.

Ethanol mole balance: $L'X_6 + G'Y_8 = L'X_7 + G'Y_7$

Ethanol equilibrium: $Y_7/(1 + Y_7) = K_7X_7/(1 + X_7)$

PS (5%): $L'X_7 = 0.05L'X_0$ (note: if 95% stripped, 5% remains)

Stage 6
Ethanol mole balance: $L'X_5 + G'Y_7 = L'X_6 + G'Y_6$

Ethanol equilibrium: $Y_6/(1 + Y_6) = K_6X_6/(1 + X_6)$

Stages 4 – 1
These are continued and include one that compares the calculated liquid feed concentration (X_0) to that specified (in this case, 0.02041 mol E/mol W). The solution is found in Excel "Stripper/7-Stage." Run Solver to change variables in order to satisfy the equations. With $G' = 232$ mol air, the compositions and K values are shown below.

Stage	X (mol E/mol W)	Y (mol E/mol A)	K
(0)	0.0204		
1	0.0164	0.0082	0.5026
2	0.0127	0.0065	0.5168
3	0.0093	0.0049	0.5300
4	0.0065	0.0035	0.5418
5	0.0042	0.0023	0.5518
6	0.0024	0.0013	0.5599
7	0.0010	0.0006	0.5662
(8)		0.0000	
	Y_8 Specified =	0.0000	
	Δ	0.0000	

Compare the 232 mol/s of air for a seven-stage stripper to 686 mol/s and 3,290 mol/s required by two- and one-stage strippers with the same specifications. Notice that the K value is significantly lower than assumed for the absorber.

d. Explore the problem.

Try removing 95% of a much higher ethanol composition (20 mol%) in the feed. Note that only the bottom stages experience much change in concentration. This means that the separation is relatively easy and does not really need seven stages. Why is the separation "easy?" Under these conditions, equilibrium strongly favors the gas phase. You will need to change the axis in order to see the stair steps.

e. Be awesome!

Explore the process conditions T and P to see how they affect the performance of the column. When you are done with this chapter (and we are close!), do something entirely different, like taking a long walk or reading a chapter in a novel.

13.6 Yet Another Interesting Observation

The passing and leaving streams from each stage of the stripper in Example 13.4 are plotted in Figure 13.8. Vertical and horizontal lines are drawn between alternating passing and leaving streams. This looks similar to Figure 13.6 for the seven-stage absorber except the passing streams are below the equilibrium line. This indicates that a graphical technique exists for strippers too.

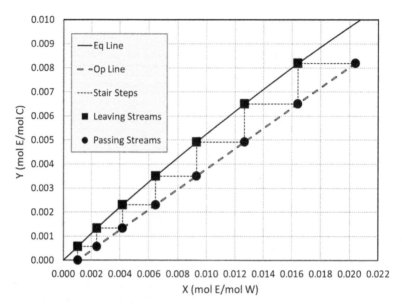

FIGURE 13.8
Demonstration of stepping off stages for the seven-stage stripper used in Example 13.4.

13.7 Summary

Better separation will occur with increasing number of stages for both absorbers and strippers. This can mean increased transfer of chemical between phases for the same flow rate or less flow rate needed for the same degree of separation. The equations used to analyze the processes are mole balances, phase equilibrium relationships, and various process specifications. Solvent-free coordinates were introduced to simplify the calculations. The passing and leaving streams between and from stages can be plotted on an XY diagram and connected to form stairs, just like the McCabe–Thiele design, leading us to "discover" a graphical method for column design.

14

Graphical Design of Absorption/ Stripping Columns

Absorption and stripping columns can be designed graphically using the McCabe–Thiele method similarly to distillation columns. In Chapter 13, we saw that plotting passing and leaving stream compositions (solute-free) from stage to stage [(X_0, Y_1); (X_1, Y_1); (X_1, Y_2); (X_2, Y_2); and so on] created stair steps from the bottom of the column (X_N, Y_{N+1}) to the top (X_0, Y_1). The passing streams are related by mole balances and lie on a single operating line if there are no intermediate feed or product streams. The leaving streams all lie on the phase equilibrium curve if we assume equilibrium stages. In this chapter, we will develop the equations for these two lines and then show how to use a graphical method to design the process.

14.1 Phase Equilibrium Curve

Each point on an XY diagram satisfies the modified Raoult's law for ethanol.

$$y = Kx \tag{14.1}$$

$$\frac{Y}{1+Y} = K\frac{X}{1+X} \tag{14.2}$$

x (mol E/mol) and y (mole E/mol) are mole fractions in the liquid and gas phases, respectively, X (mol E/mol W) and Y (mole E/mol C) are mole ratios, and K is the vapor-liquid equilibrium (VLE) constant. K is calculated using the Wilson equation for the activity coefficient.

$$K = \frac{y}{x} = \gamma_E P_E^* / P \tag{14.3}$$

Details about this calculation are found in Chapter 11. Purely to demonstrate calculations, in Chapters 12 and 13 we assumed an ethanol activity coefficient, $\gamma_E = 6.0$ at infinite dilution, for absorbers and calculated it using the Wilson equation for strippers. Ethanol phase equilibrium is calculated for a range of values of X to find the associated Y at 30°C and 825 mm Hg in Excel "Graphical/Equilibrium" and plotted in Figure 14.1. For comparison,

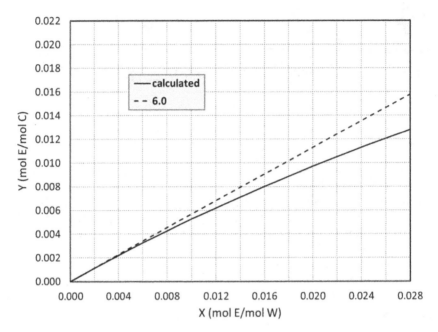

FIGURE 14.1

Vapor–liquid equilibrium in solute-free coordinates for ethanol in water and carbon dioxide at 30°C and 825 mm Hg. The dotted line is the result of assuming $\gamma_E = 6.0$ (infinite dilution) and the solid line is the result of calculating γ_E using the Wilson equation. It makes a difference!

the VLE curve using a constant activity coefficient of $\gamma_E = 6.0$ is also plotted. It looks like infinite dilution is not a very good assumption, even at these low concentrations. Let's use the full Wilson equation in all future examples using ethanol and water.

14.2 Absorber Operating Line

The operating line relates the passing streams between stages using a mole balance. Referring to the N-stage column in Figure 14.2, the mole balance around Stage 1 requires that the moles of ethanol entering with the liquid and gas flows must equal the moles of ethanol leaving with the liquid and gas flows.

$$L'X_0 + G'Y_2 = L'X_1 + G'Y_1 \tag{14.4}$$

L' (mol E/mol W) and G' (mol E/mol C) are solute-free liquid and gas molar flow rates, respectively. Rearranging Eq. 14.4 in the form of a straight-line equation relating the composition of passing streams (X_1, Y_2) yields

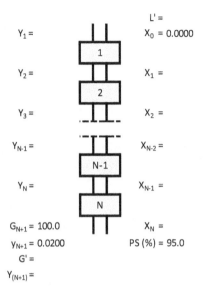

FIGURE 14.2
Process flow diagram for the absorber with an unknown number of stages. It is used in developing the equation for an operating line and is described in Example 14.1.

$$Y_2 = \frac{L'}{G'} X_1 + Y_1 - \frac{L'}{G'} X_0 \tag{14.5}$$

With the material balance envelope drawn around Stages 1 and 2, the mole balance is

$$Y_3 = \frac{L'}{G'} X_2 + Y_1 - \frac{L'}{G'} X_0 \tag{14.6}$$

More generally, a material balance envelope around all stages from Stage 1 to i is

$$Y_{i+1} = \frac{L'}{G'} X_i + Y_1 - \frac{L'}{G'} X_0 \tag{14.7}$$

Understanding X and Y to be passing streams (X_i, Y_{i+1}) and assuming that the absorbent feed has no solute, this can be simplified to

$$Y = \frac{L'}{G'} X + Y_1 \tag{14.8}$$

To draw the operating line, either a point and the slope or two points are needed. Notice how similar this development is to that for distillation columns.

Analytical solutions predicted the performance of a column – given the number of equilibrium stages, how much solute can be transferred between phases? Graphical solutions provide a design of the column – how many stages are needed for a certain performance? Now that both the equilibrium and operating lines can be found, let's try a couple examples.

**EXAMPLE 14.1   Absorption Column
with Unknown Number of Stages**

A total of 100 mol/s of gas containing 2.0 mol% ethanol (E) and balance carbon dioxide (C) is contacted with pure water (W) in an absorber operating at 30°C and 825 mm Hg. About 95 mol% of the ethanol is absorbed. Determine the flow rates and compositions of all streams and the number of stages by graphical methods using solute-free mole fractions and flow rates.

a. Draw a labeled process flow diagram.
   The labeled PFD was already shown in Figure 14.2.
b. Conduct a degree of freedom analysis.
   Can this be solved? This is similar to the problem statement in Example 13.3 except the number of stages is also unknown. Do we have enough information to find two points for the operating line? The top mole ratios (X_0, Y_1) are (0.0000, 0.0010) as determined from an overall mole balance with the 95% removal specification.

$$Y_1 = \frac{(2 \text{ mol E})}{98 \text{ mol C}}(1 - 0.95) = 0.0010 \text{ mol E/mol C}$$

However, since we don't know L′, we can't calculate the bottom liquid mole ratio X_N or the slope, L′/G′. Comparing this to Example 13.3, we added an extra unknown (number of stages) and this means we will need one more piece of information.
   We will additionally specify that 61.7 mol/s of water is fed to the top of the column. I know from the analytical solution in Excel "Graphical/Absorber" that this water flow rate is required to remove 95% of the ethanol using seven equilibrium stages. So if we do this right, we should find that exactly seven stages are required using this graphical method. (Compare this to 69.5 mol W/s using K = 0.57.)

c. Write the equations and solve.
   For a graphical solution, we need the equilibrium and operating lines. The equilibrium line is determined by the temperature and pressure. See Excel "Graphical/VLE" that can be

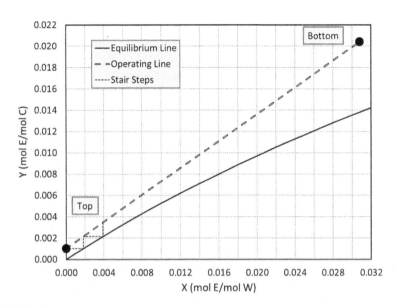

FIGURE 14.3
The first steps between the equilibrium and operating line for Example 14.1.

updated with a change in T and P. The XY graph is shown in Figure 14.3 with some additional features to be explained.

The operating line can be drawn with one passing stream point and the slope (L'/G' ratio) or two passing stream points. The top point was $(X_0, Y_1) = (0.0000, 0.0010)$ and the bottom point is $(X_N, Y_{N+1}) = (0.0308, 0.0204)$ as determined from an overall mole balance.

$$X_N = \frac{(2 \text{ mol E})(0.95)}{61.7 \text{ mol W}} = 0.0308 \text{ mol E/mol W}$$

We could also use the top point and the slope.

$$\frac{L'}{G'} = \frac{61.7 \text{ mol W}}{98 \text{ mol C}} = 0.630 \text{ mol W/mol C}$$

However, it is easier to use the two points. These are drawn on Figure 14.3 with a connecting straight line (the operating line). As a double check, the slope of that line is

$$\text{slope} = \frac{\text{rise}}{\text{run}} = \frac{0.0204 - 0.0010}{0.0308 - 0} = 0.630$$

This agrees with the above calculation of slope.

The graphical method can start at the top or bottom of the column. Starting at the top of the column (this might seem confusing, but this is at the lower left part of Figure 14.3) locate the passing stream composition $(X_0, Y_1) = (0.0000, 0.0010)$. Next, find the composition of the liquid that is in equilibrium with the gas leaving Stage 1 by moving horizontally to the equilibrium line $(X_1, Y_1) = (0.0018, 0.0010)$. Then, find the composition of the gas stream passing between Stages 1 and 2 by moving vertically to the operating line $(X_1, Y_2) = (0.0018, 0.0022)$.

Would you like to see this done for one more stage? Find the composition of the liquid that is in equilibrium with the gas leaving Stage 2 by moving horizontally to the equilibrium line $(X_2, Y_2) = (0.0039, 0.0022)$. Then, find the composition of the gas stream passing between Stages 2 and 3 by moving vertically to the operating line $(X_2, Y_3) = (0.0039, 0.0035)$.

Continue moving vertically and horizontally and noting the leaving (equilibrium curve) and passing stream (operating line) compositions. The first steps are shown in Figure 14.3. Complete the process using a straight edge and pencil. You should find that seven stages are required and the result looks exactly like Figure 14.4. Graphical techniques like this depend on good eyesight, a steady hand, and patience. Don't be surprised if you

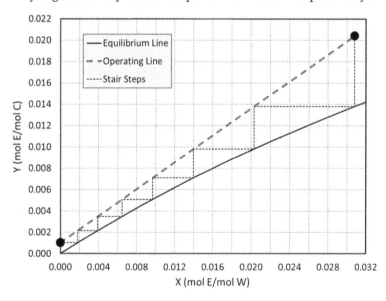

FIGURE 14.4
Completed graphical solution for the absorber in Example 14.1 showing exactly seven stages required.

don't "land" exactly on the bottom passing stream compositions. In the case of arbitrary specifications, it would be pure
coincidence if you didn't overshoot your goal, just as you did
with the McCabe–Thiele method in distillation.

d. Explore the problem.

After you very carefully and patiently complete the process
on Figure 14.3, print another copy of the equilibrium graph from
Excel "Graphical/XY" and repeat the process freehand and as fast
as you can. The result is probably a bit different, isn't it! But it still
serves as a quick check and might reveal problems with the specifications, for instance, asking the process to do the impossible, like
using less than the minimum absorbent flow rate. Speaking of …

e. Be awesome!

What happens if you use less absorbent? Change the X-axis
of the equilibrium graph in Excel "Graphical/XY" to 0.045 mol
E/mol W, print the graph, and repeat this problem for 50 mol/s
of water. This requires more stages, doesn't it? Also, the liquid
product composition increases because the same amount of
ethanol is in less water. What happens as you keep decreasing the amount of water? Is there a minimum amount of water
that can achieve 95% absorption with any number of stages
(infinite)? Figure 14.5 shows the operating lines for 61.7, 50.0,

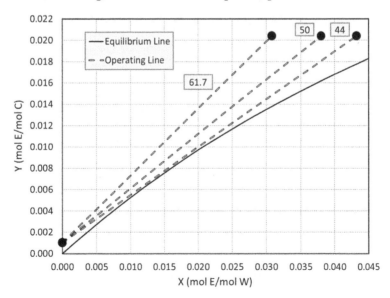

FIGURE 14.5
Operating lines for various water flow rates (mol/s) for Example 14.1e.

and 44.0 mol W/s. Try to step off the stages starting at the top of the column (remember – this is at the lower left corner of the graph). Don't try too hard for 44.0 mol W/s – you have better things to do! Pinch points where the operating line intersects the equilibrium line require an infinite number of stages.

14.3 Minimum Absorbent Flow Rate

In part (e) of Example 14.1, we found that there appears to be a minimum flow rate that will achieve a certain performance no matter how many stages are required. This happens when the operating line intersects the equilibrium line. That can happen at (X_N, Y_{N+1}) or somewhere in between depending on the shape of the equilibrium curve, as we will see in the next example. The minimum flow rate can be found analytically by solving the operating line and equilibrium equation simultaneously. However, it is more easily found graphically by using the slope of the resulting operating line to find L′. This will be demonstrated in the next example.

EXAMPLE 14.2   Multi-Stage Absorber with Minimum Absorbent Flow Rate Specified

A total of 100 mol/s of gas containing 2.0 mol% ethanol (E) and balance carbon dioxide (C) is contacted with pure water (W) in an absorber operating at 30°C and 825 mm Hg. About 150% of the minimum water flow rate is used. Determine the flow rates and compositions of all streams and the number of stages required.

A common process specification is to call for an absorbent flow rate that is some percentage above the minimum. A typical number is in the ballpark of 150% above the minimum.

a. Draw a labeled process flow diagram.
    The PFD is shown in Figure 14.6. The inlet gas stream continues to be G′ = 98.0 mol C/s and Y_{N+1} = 0.0204 mol E/mol C. Add those values to the figure.
b. Conduct a degree of freedom analysis.
    It seems like we don't have enough information. We don't know the minimum flow rate or the amount absorbed. Let's check.

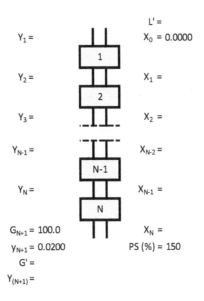

$Y_1 =$

$L' =$
$X_0 = 0.0000$

$Y_2 =$ $X_1 =$

$Y_3 =$ $X_2 =$

$Y_{N-1} =$ $X_{N-2} =$

$Y_N =$ $X_{N-1} =$

$G_{N+1} = 100.0$ $X_N =$
$y_{N+1} = 0.0200$ PS (%) = 150
$G' =$
$Y_{(N+1)} =$

FIGURE 14.6
PFD of the multi-stage absorber described in Example 14.2.

Stage N

Unknowns (4): L', X_N, X_{N+1}, Y_N

Equations (3):

　　1 mole balance (ethanol)

　　1 equilibrium relationship (ethanol)

　　1 process specification (150% of minimum)

We need one more piece of information. Let's specify 80% absorption of ethanol and continue.

c. Write the equations and solve.

　　The first thing we need is the minimum flow rate. If 80% of the ethanol is absorbed, then the product gas composition is

$$Y_1 = \frac{(1-0.8)2.0}{98} = 0.0041$$

The operating line will then start at (0.0000, 0.0041). The intersection of the operating line and the equilibrium line – where a pinch point would occur – happens at the coordinates $(X_N, 0.0204)$ as seen in Figure 14.7.

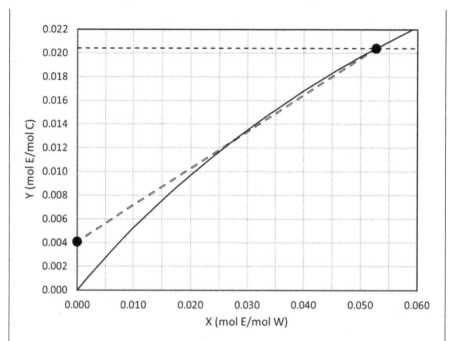

FIGURE 14.7

Location of operating line with bottom passing streams at Y_{N+1} on the equilibrium curve.

But wait a minute! The operating line crosses the equilibrium line at ~(0.027, 0.012) and certainly approaching this point from the top of the column would be a pinch point. But a pinch point would occur at a larger value for L′ (steeper slope) if the operating line were just tangent to the equilibrium curve as done in Figure 14.8. Approaching a minimum L′ from a larger flow rate would encounter the tangent point first.

The slope is

$$\frac{L'}{G'} = \frac{0.0204 - 0.0041}{0.0513 - 0} = 0.318$$

And the minimum liquid flow rate is

$$L'_{min} = (0.318)(98) = 31.2 \text{ mol W}$$

If you tried to step off the stages, you would find a "pinch point" near the tangent point requiring an infinite number of stages.

The specified liquid flow rate is 150% above the minimum.

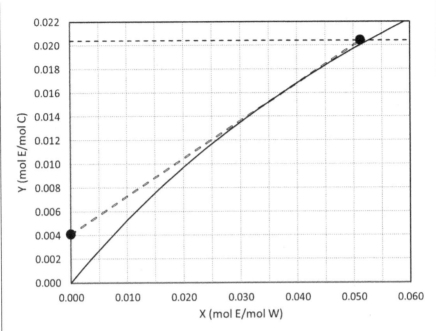

FIGURE 14.8
Operating line for minimum absorbent flow rate for the absorber specified in Example 14.2.

$$L' = (1.5)(31.2) = 46.7 \text{ mol}$$

The operating line at this flow rate is

$$Y = (L'/G')X + Y_1 = (46.7/98)X + 0.0041 = 0.477X + 0.0041$$

This is drawn in Figure 14.9. The stages are stepped off from the top of the column and we find that four stages significantly overshoots the specification.

d. Explore the problem.

Decrease the water flow rate until four stages just satisfies the specification of 80% removal. This requires some trial and error. Print some additional XY graphs from Excel "Graphical/XY."

e. Be awesome!

Modify Excel "Absorber/7-Stage" to solve for a four-stage column to meet the same specifications. Compare your analytical solution to your graphical solution above.

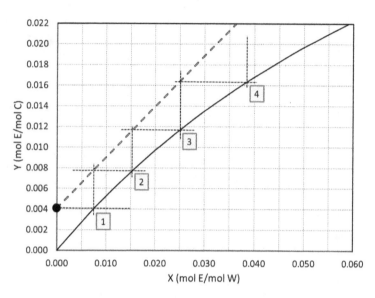

FIGURE 14.9
McCabe–Thiele design of the absorber described in Example 14.2.

14.4 Stripping Operating Line

The mole balance equation for absorption passing streams (Eq. 14.7) is also valid for a stripping column.

$$Y_{i+1} = \frac{L'}{G'}X_i + Y_1 - \frac{L'}{G'}X_0 \qquad (14.7)$$

We can't eliminate the last term as we did for absorbers. Another complication is that the Y-intercept is negative since the operating line is below the equilibrium curve. However, it shows that we can still draw the operating line with two points or one point and the slope.

EXAMPLE 14.3   Stripping Column with Unknown Number of Stages

A total of 100 mol/s of a liquid containing 2.0 mol% ethanol in water is contacted with 232 mol of air in an equilibrium-stage column. Find the composition and flow rate of all streams and the number of stages.

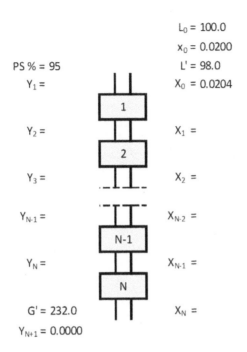

PS % = 95
$Y_1 =$

$Y_2 =$

$Y_3 =$

$Y_{N-1} =$

$Y_N =$

G' = 232.0
$Y_{N+1} = 0.0000$

$L_0 = 100.0$
$x_0 = 0.0200$
L' = 98.0
$X_0 = 0.0204$

$X_1 =$

$X_2 =$

$X_{N-2} =$

$X_{N-1} =$

$X_N =$

FIGURE 14.10
PFD of the stripping column described in Example 14.3.

We repeated Try This at Home 13.2 here so that we should find exactly seven stages that are required.

  a. Draw a labeled process flow diagram.
    The PFD is shown in Figure 14.10. The feed stream has L' = 98 mol W/s and $X_0 = 0.0204$ mol E/mol W.
  b. Conduct a degree of freedom analysis.

Stage N

Unknowns (3): Y_N, X_{N-1}, X_N

Equations (3):
    1 mole balance (ethanol)
    1 equilibrium relationship (ethanol)
    1 process specification (95% stripped)

There are zero DOF and the unknowns in Stage N can be determined. This will allow the remaining stages to be solved sequentially until X_0 reaches 0.0204 mol E/mol W at the unknown top stage.

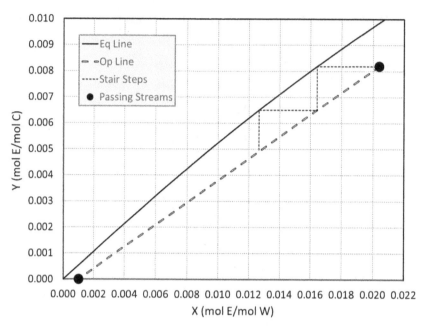

FIGURE 14.11

Operating line for the stripping column described in Example 14.3 with the first few McCabe–Thiele steps constructed.

c. Write the equations and solve.

This will be done graphically, but first we must find X_N and Y_1 from the process specification.

$$X_N = (0.05)(2.00)/98.0 = 0.0010$$

$$Y_1 = (0.95)(2.00)/232.0 = 0.0082$$

The passing streams at the bottom of the column are then $(X_N, Y_{N+1}) = (0.0010, 0.0000)$ and at the top of the column are $(X_0, Y_1) = (0.0204, 0.0082)$. These are marked on Figure 14.11 and connected by the operating line.

Note that the operating line for stripping lies below the equilibrium line and that the bottom of the column is at the bottom left of the graph. Starting at the top of the column $(X_0, Y_1) = (0.0204, 0.0082)$, find the liquid stream in equilibrium with the gas leaving Stage 1 by drawing a horizontal line to the equilibrium line at $(X_1, Y_1) = (0.0164, 0.0082)$. Now find the composition of the gas stream passing between Stages 1 and 2 by drawing a vertical line to the operating line at $(X_1, Y_2) =$ (0.0164, 0.0065). Repeating these steps a few more times, draw a horizontal line to the equilibrium line at $(X_2, Y_2) = (0.0127,$ 0.0065) and then a vertical line to the operating line at $(X_2, Y_3) =$ (0.0127, 0.0049). These first steps are shown on Figure 14.11. Carefully continue this process until you reach the operating line at the bottom of the column at $(X_N, Y_{N+1}) = (0.0010, 0.0000)$. Hopefully, you "landed" on this point after seven stages. Your graph should look very similar to Figure 13.8 because we used the gas flow rate of 232 mol from the analytical solution for Try This at Home 13.2. Read off all compositions from the graph and fill in the PFD in Figure 14.10.

d. Explore the problem.

Decrease the amount of stripping gas. What does this do to the slope of the operating line? Show that it moves closer to the equilibrium line and results in more stages to meet the specifications. Increase the amount of stripping gas. By trial and error, find the gas flow rate required to meet these specifications with four stages.

e. Be awesome!

Find the minimum gas flow rate that meets these specifications with an infinite number of stages. Show that $G' = 232\,mol$ air is ~120% of the minimum flow rate.

TRY THIS AT HOME 14.1   Multi-Stage Stripper with Minimum Gas Flow Rate Specified

A total of 100 mol/s of a liquid stream containing 5 mol% ethanol in water is contacted with dry air in a stripping column. The airflow rate is 150% of the minimum. Enough stages are present to remove 90% of the ethanol. Find the number of stages and the flow rate and composition of all streams.

a. Draw a labeled process flow diagram.

The PFD is drawn in Figure 14.12. Fill in what you know from the problem statement. Determine L' and X_0 and add those values. Leave the other variables blank for now.

b. Conduct a degree of freedom analysis.

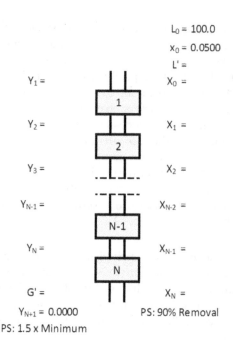

FIGURE 14.12
PFD of the stripping column described in Try This at Home 14.1.

Stage 1

Unknowns (4):

Equations (4):

> There is sufficient information to solve for the airflow rate and all stream compositions for Stage 1. The problem can then be solved sequentially down the column.

c. Write the equations and solve.
   Determine the composition of the liquid product. Define $f \equiv$ fraction removed. Pencil in your calculation.

$$X_N = X_0(1-f) = \qquad = 0.0053$$

An XY graph is shown in Figure 14.13. Locate the passing stream compositions at the bottom of the column (0.0053, 0.0000) and draw a line from there to the intersection of the equilibrium curve at $X_0 = 0.0526$. Y at this point should be about 0.0204. This is the operating line with minimum airflow rate. The pinch point at (0.053, 0.020) will require an infinite number of stages to achieve this performance.

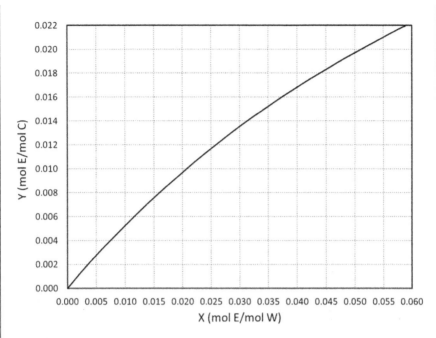

FIGURE 14.13
XY diagram for Try This at Home 14.1.

Because of the equilibrium curve shape, the operating line is not tangent to the curve at any point so the intersection is used.

The slope of this line is

$$\frac{\text{rise}}{\text{run}} = \frac{(0.0204 - 0)}{(0.0526 - 0.0053)} = 0.431$$

The minimum airflow rate is

$$\frac{L'}{G'} = \frac{95}{G'} = 0.431 \qquad G' = \frac{95}{0.431} = 220 \text{ mol/s}$$

150% of the minimum airflow rate is then

$$G' = (1.5)(220) = 330 \text{ mol/s}$$

At the top of the column, we then have

$$\frac{L'}{G'} = \frac{95}{330} = 2.87 = \frac{\text{rise}}{\text{run}} = \frac{(Y_1 - 0)}{(0.0526 - 0.0053)} \qquad Y_1 = 0.0136$$

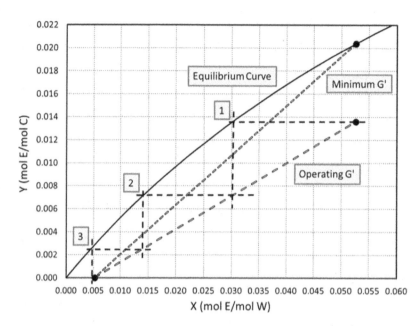

FIGURE 14.14
Completed McCabe–Thiele analysis for Try This at Home 14.1.

To start a graphical solution, locate the passing stream compositions at the top (X_0, Y_1) and bottom (X_N, Y_{N+1}) of the column and connect them with a straight operating line. Proceed with constructing stair steps from the top to the bottom. You should find that three stages are just enough to meet the 90% removal specification. Modify Figure 14.12 appropriately and fill in all flow rates and compositions.

My graph with all work is shown in Figure 14.14. Does yours look something like this?

d. Explore the problem.

Print some XY graphs and try several airflow rates to see how it affects the number of stages.

e. Be awesome!

Modify Excel "Stripping/7-Stage" to make it three stages. Use the specifications of this problem to see if your analytical results match the graphical solution.

14.5 Summary

Using a mole balance around the top of a column and any number of stages, we found that an operating line could be described by $Y = (L'/G') X + Y1 - (L'/G') X0$, where X and Y are passing streams between stages. The operating line is above the equilibrium curve for absorption and below for stripping. A stair step can be constructed to graphically determine the passing and leaving stream compositions. Each pair of leaving stream compositions (Xi, Yi) represents a stage. When we used the analytical solutions for a seven-stage column from Chapter 13, the graphical solutions matched, as expected.

15

Multicomponent and Adiabatic Absorption Columns

For a three-component system, ethanol being absorbed from a carbon dioxide gas stream into water, we assumed that only the solute transferred between phases and that the system was isothermal. How good were these assumptions?

In this chapter, we will focus on the absorber. We know that water will indeed evaporate into the gas stream. In fact, some carbon dioxide should dissolve into the liquid stream along with the ethanol to produce a carbonated alcoholic beverage ... uh, product. Water evaporation should cool the system (this is called evaporative cooling) and the absorption of ethanol and carbon dioxide should heat the system. Will these differences significantly change our analysis?

Because all three components are changing phases, it no longer makes sense to use the solute-free coordinates and two-dimensional vapor-liquid equilibrium (VLE) graphs. Instead, we'll go back to using molar flow rates with mole fractions, and just use mathematical methods.

15.1 Thermodynamics

Let's assume that the vapor and liquid leaving each stage will be in equilibrium for all three components. Just like ethanol equilibrium, we'll use the modified Raoult's law for water and Wilson's equation to determine the activity coefficient. The carbon dioxide will dissolve at low concentrations that should follow Henry's law. The Henry's law constant for carbon dioxide in water is given by

$$\ln H = -1,392/T + 0.1210 \ln T + 0.01479T + 3.984 \tag{15.1}$$

where H [=] psia/mol fraction and T [=] R. It can easily be converted to pressure units mm Hg. Here we make two important assumptions. The activity coefficients for ethanol and water are not affected by the small amount of carbon dioxide in solution. Also, the Henry's law constant for carbon dioxide in a water/ethanol mixture is the same as in pure water.

Instead of assuming an isothermal column, we can assume it is adiabatic so that the steady-state enthalpy balance is

$$Q \equiv 0 = \Delta H = \sum_{out} n_i \hat{H}_i - \sum_{in} n_i \hat{H}_i \qquad (15.2)$$

where Q is the heat transferred (kJ/s), H is the total enthalpy (kJ/s), $\hat{H}$ is the molar enthalpy of each species (kJ/mol), and n is the molar flow rate of each species (mol/s). I will use the feed conditions of all species as the reference state: water (liquid, 30°C), ethanol (vapor, 30°C), and carbon dioxide (gas, 30°C) and 825 mm Hg. Therefore, the enthalpy of all feed streams will be 0 kJ/mol.

The enthalpy of a stream, H, is the product of the specific enthalpy of each species at the stream temperature and phase, the molar flow rate, and the composition. The gas stream enthalpy, for example, is

$$H_G = G\left(y_W \hat{H}_{WG} + y_C \hat{H}_{CG} + y_E \hat{H}_{EG} \right) \qquad (15.3)$$

where G is the total gas flow rate (mol/s), y is the mol fraction, and $\hat{H}$ is the specific enthalpy (kJ/mol).

The reference state for carbon dioxide and ethanol is gas, so only sensible heat is required to bring them to the system temperature from the feed temperature. The hypothetical path for carbon dioxide in the gas stream is C (gas, T_F) → C (gas, T) and the same for ethanol. The enthalpy of carbon dioxide is calculated by integrating the heat capacity at constant pressure, C_{PV} (kJ/mol/°C), from the feed temperature, T_F (°C), to the system temperature, T (°C).

$$\hat{H}_{CG} = \int_{T_F}^{T} C_{PV} \, dT \qquad (15.4)$$

A similar calculation can be made for the ethanol.

Since the water's reference state is a liquid, the hypothetical path to calculate its enthalpy in the gas includes latent heat and is W (liquid, T_F) → W (liquid, T_B) → W (vapor, T_B) → W (vapor, T).

$$\hat{H}_{WV} = \int_{T_F}^{T_B} C_{PL} \, dT + \Delta \hat{H}_V + \int_{T_B}^{T} C_{PV} \, dT \qquad (15.5)$$

T_B is the normal boiling point (°C), C_{PV} is the vapor-phase heat capacity at constant pressure, C_{PL} is the liquid-phase heat capacity at constant pressure (both in units kJ/mol/°C), and $\Delta \hat{H}_V$ is the heat of vaporization (kJ/mol), usually given at the normal boiling temperature.

The enthalpy of the liquid product is also the sum of the component enthalpies

$$H_L = L\left(x_W \hat{H}_{WL} + x_C \hat{H}_{CL} + x_E \hat{H}_{EL}\right) \tag{15.6}$$

where L is the liquid molar flow rate (mol/s) x is the mol fraction, and $\hat{H}$ is the specific enthalpy (kJ/mol).

The reference state for the water is liquid at 30°C and 825 mm Hg so only sensible heat is required to bring it to system temperature in the liquid stream. The hypothetical path is W (liquid, T_F) → W (liquid, T).

$$\hat{H}_{WL} = \int_{T_F}^{T} C_{PL}\, dT \tag{15.7}$$

The reference state for the ethanol and carbon dioxide is vapor/gas at 30°C so their enthalpies in the liquid will include both sensible and latent heat. The hypothetical path for ethanol is E (vapor, T_F) → (vapor, T_B) → E (liquid, T_B) → E (liquid, T).

$$\hat{H}_{EL} = \int_{T_F}^{T_B} C_{PV}\, dT - \Delta\hat{H}_V + \int_{T_B}^{T} C_{PL}\, dT \tag{15.8}$$

The calculation will be similar for carbon dioxide except it will be heat of solution instead of heat of vaporization. For lack of contrary information, we'll assume the heat of solution is not a function of temperature, at least in the range of temperatures here. So it is simply

$$\hat{H}_{CL} = \int_{T_F}^{T} C_P\, dT + \Delta\hat{H}_S \tag{15.9}$$

15.2 Rigorous Analysis

Using the energy balance instead of the equimolal overflow (EMO) assumption in distillation was considered a rigorous method. In the same way, we use the energy balance here rather than simply assume isothermal operation. It is best to demonstrate this for one- and two-stage absorbers. The results from a process simulator (ChemCAD) analysis of a seven-stage absorber are used for comparison.

EXAMPLE 15.1   Absorption in an Adiabatic Single Stage

A total of 100 mol/s of 2.0 mol% ethanol in carbon dioxide is fed to a single equilibrium stage at 30°C and 825 mm Hg. The gas is contacted with 200 mol/s of pure water at the same temperature and pressure to remove the ethanol. Assume the process is adiabatic (not isothermal) and that all products are in phase equilibrium.

a. Draw a labeled process flow diagram.
   A labeled PFD is shown in Figure 15.1.
b. Conduct a degree of freedom analysis.

   Unknowns (9): G, y_W, y_E, y_C, L, x_W, x_E, x_C, T

   Equations (9):
   > 3 mole balances (total, water, ethanol)
   > 3 equilibrium relationships (water, ethanol, carbon dioxide)
   > 1 energy balance (adiabatic)
   > 2 summations (liquid and gas)

   There are zero DOF and the problem can be solved.
c. Write the equations and solve.

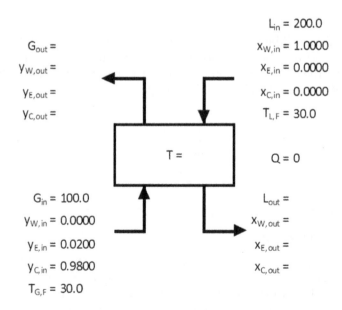

$$L_{in} = 200.0$$

$G_{out} =$

$y_{W,out} =$

$y_{E,out} =$

$y_{C,out} =$

$x_{W,in} = 1.0000$

$x_{E,in} = 0.0000$

$x_{C,in} = 0.0000$

$T_{L,F} = 30.0$

$T =$

$Q = 0$

$G_{in} = 100.0$

$y_{W,in} = 0.0000$

$y_{E,in} = 0.0200$

$y_{C,in} = 0.9800$

$T_{G,F} = 30.0$

$L_{out} =$

$x_{W,out} =$

$x_{E,out} =$

$x_{C,out} =$

FIGURE 15.1
PFD of a single-stage absorber as described in Example 15.1. G, L [=] mol/s; x, y [=] mole fraction; T [=] °C; Q [=] kJ/s, P = 825 mm Hg.

Total mole balance: $G_{in} + L_{in} = G_{out} + L_{out}$

Water mole balance: $L_{in}x_{W,in} = G_{out}y_{W,out} + L_{out}x_{W,out}$

Ethanol mole balance: $G_{in}y_{E,in} = G_{out}y_{E,out} + L_{out}x_{E,out}$

Water equilibrium: $y_{W,out} = K_W x_{W,out}$ (Note: $K_i = \gamma_i P_i^* / P$

Ethanol equilibrium: $y_{E,out} = K_E x_{E,out}$

Carbon dioxide equilibrium: $y_{C,out} = K_C x_{C,out}$ (Note: $K_C = H_C/P$)

Energy balance: $Q \equiv 0 = \Delta H = \sum_{gas} n_i \hat{H}_i + \sum_{liq} n_i \hat{H}_i$

Gas summation: $\Sigma y_i = 1$

Liquid summation: $\Sigma x_i = 1$

The enthalpy calculations are best organized in an enthalpy table as shown in Table 15.1 and the calculations are performed in Excel "Absorber EB/1-Stage" using Eqs. 15.2–15.9. The enthalpies of all feed streams are zero because that was the reference state chosen. The enthalpies of the water in the liquid phase and of the ethanol and carbon dioxide in the vapor phase are just due to the temperature change. The enthalpies of the water in the vapor phase and the ethanol and carbon dioxide in the liquid phase also included a phase change.

The temperature decreased from 30°C to 26.8°C because of evaporative cooling. The big difference here is that 1.6% of the water actually evaporated into the gas stream. Only 0.12% of the carbon dioxide dissolved in the water because of this low pressure (carbonation of water usually is done at ~2.4

TABLE 15.1

Enthalpy Table for Example 15.1

	Ref: Feed Streams					
Species	n_{in} (mol)	$\hat{H}_{in}$ (kJ/mol)	H_{in} (kJ)	n_{out} (mol)	$\hat{H}_{out}$ (kJ/mol)	H_{out} (kJ)
W(L)	200	0.00	0.00	196.8	−0.24	−47.08
E(L)	---			1.6	−44.12	−71.84
C(L)	---			0.1	−19.52	−2.31
W(V)	---			3.2	−41.38	−132.90
E(V)	2	0.00	0.00	0.4	−0.21	−0.08
C(V)	98	0.00	0.00	97.9	−0.12	−11.58
Sum			0.00			0.00

atm). About 81% of the ethanol was absorbed. The simplified method predicted absorption of 79% because the temperature is assumed constant at 30°C. More of the ethanol vapor will absorb at a lower temperature.

d. Explore the problem.

How can we remove 95 mol% of the ethanol from the gas? Would you be surprised that there are at least three different process variables to use? Try increasing the amount of water to 1,040 mol/s. Decrease the feed temperature to 1.1°C. At a lower temperature, the equilibrium shifts to favor the liquid phase. Notice that the products leave at a higher temperature, 4.3°C, because of the heat of condensation as the vapor phase ethanol joins the liquid phase. How about changing the pressure to 4,863 mm Hg or over 6 atm? Here the temperature also increases from 30°C to 33°C because of the increased amount of ethanol condensing. Amaze your friends with this great party trick!

For each change in process conditions, verify by hand calculations that the equations are satisfied and make sure you can explain the effects to your study partners.

e. Be awesome!

As always, a DIY version of the spreadsheet is attached. Input the equations and verify that you get the same solution.

EXAMPLE 15.2   Absorption in Two Adiabatic Stages

The same conditions as Example 15.1 apply except that it will be a two-stage column. The same equations will be used but there will be twice as many.

a. Draw a labeled process flow diagram.

The PFD is shown in Figure 15.2.

b. Conduct a degree of freedom analysis.

Look at the complete system with both stages and all streams.

Unknowns (18): G_1, y_{W1}, y_{E1}, y_{C1}, G_2, y_{W2}, y_{E2}, y_{C2}, T_1, L_1, x_{W1}, x_{E1}, x_{C1}, L_2, x_{W2}, x_{E2}, x_{C2}, T_2

Equations (18):

For each stage:

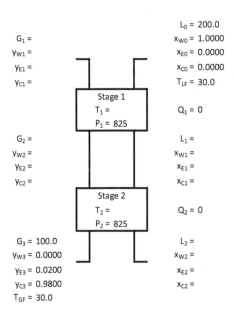

FIGURE 15.2

PFD of the two-stage absorber described in Example 15.2. G, L [=] mol/s; x, y [=] mole fraction; T [=] °C; Q [=] kJ/s, P [=] mm Hg. Subscripts 0 and 3 refer to liquid and gas feeds, respectively.

> 3 mole balances (total, water, ethanol)
>
> 3 equilibrium relationships (water, ethanol, carbon dioxide)
>
> 2 summations (liquid and vapor)
>
> 1 energy balance

There are zero DOF and the problem can be solved.

c. Write the equations and solve.

For Stage 1 the liquid feed flow rate, L_0, and mole fractions are known.

Total mole balance: $L_0 + G_2 = L_1 + G_1$

Water mole balance: $L_0 x_{W0} + G_2 y_{W2} = L_1 x_{W1} + G_1 y_{W1}$

Ethanol mole balance: $L_0 x_{E0} + G_2 y_{E2} = L_1 x_{E1} + G_1 y_{E1}$

Water equilibrium: $y_{W1} = K_{W1} x_{W1}$ (Note: $K_i = \gamma_i P_i^* / P$)

Ethanol equilibrium: $y_{E1} = K_{E1} x_{E1}$

Carbon dioxide equilibrium: $y_{C1} = K_{C1} x_{C1}$ (Note: $K_C = H_C / P$)

Energy balance: $Q \equiv 0 = \Delta H = \sum_{gas} n_i \hat{H}_i + \sum_{liq} n_i \hat{H}_i$

For Stage 2 the gas feed flow rate, G_3, and mole fractions are known. The equations will be equivalent.

These are solved with the "brute force" method in Excel "Absorber EB/2-Stage." With two stages, the recovery of ethanol is 95.6%, an increase from 81.4% in a single stage. A comparison of the one- and two-stage absorbers is made in Table 15.2. Also in the table are the values obtained from ChemCAD, which are pretty close considering the assumptions we made. If you use a process simulator, be sure to use the Wilson equation for water and ethanol and Henry's law for carbon dioxide.

d. Explore the problem.

Try playing with the inlet temperature, water flow rate, pressure, concentrations, etc., to observe their effects. Compare to a process simulator and explain the effects to your study partners.

e. Be awesome!

As always, a DIY version of the two-stage absorber is provided. See if you can input all the equations and reproduce the results. If you can think of a better – perhaps more flexible – way to build the simulation, by all means start from scratch!

TABLE 15.2

One-Stage and Two-Stage Absorber Results

	One-Stage Results								
	T	L	x_W	x_E	x_C	G	y_W	y_E	y_C
Excel	26.8	198.5	0.991	0.008	0.001	101.5	0.032	0.004	0.965
ChemCAD	26.3	198.6	0.992	0.008	0.001	101.4	0.031	0.004	0.965

	Two-Stage Results								
ChemCAD Stage	T	L	x_W	x_E	x_C	G	y_W	y_E	y_C
1	28.4	199.8	0.998	0.002	0.001	101.57	0.035	0.001	0.964
2	25.3	198.4	0.990	0.009	0.001	101.41	0.029	0.005	0.966
Excel Stage	T	L	x_W	x_E	x_C	G	y_W	y_E	y_C
1	28.1	200.2	0.998	0.002	0.001	101.4	0.034	0.001	0.965
2	26.6	198.6	0.990	0.010	0.001	101.6	0.031	0.004	0.965

EXAMPLE 15.3   Absorption in Seven Adiabatic Stages

The same conditions as Example 15.1 apply except that it will be a seven-stage column. The number of equations and the complexity of the non-ideal ethanol/water system start to make this a difficult project to program in Excel. Advanced matrix methods can be used to solve such systems, but will not be covered here. Instead, let's just use a process simulator and compare the results to the "simple" solution in Chapter 13 in which we assumed isothermal operation and no phase change for water and carbon dioxide.

The stage-by-stage results are shown in Figures 15.3 (temperature profile), Figure 15.4 (liquid composition profiles), and Figure 15.5 (gas composition profiles). The composition profiles are in some ways easier to understand as total molar flow rate obtained by multiplying the molar flow rate by the mole fraction and these are shown in Table 15.3.

The temperature continually decreases down the column from 29.8°C to 24.4°C as more water evaporates into the gas. Ethanol condensing into the water releases heat and moderates the cooling. A ChemCAD simulation with pure carbon dioxide and all other variables

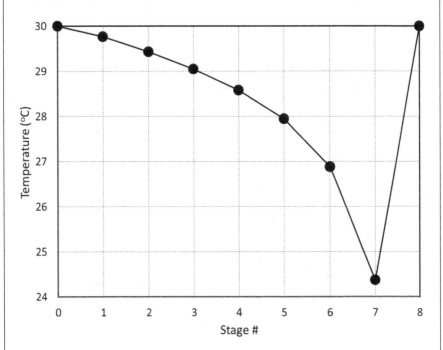

FIGURE 15.3
Temperature profile for the seven-stage absorption column in Example 15.3. Data is from ChemCAD. The simple method outlined in Chapter 13 just assumed a constant 30°C.

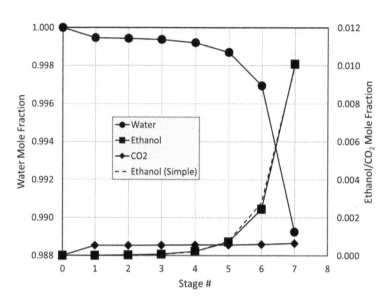

FIGURE 15.4
Liquid composition profiles for the seven-stage absorption column in Example 15.3. Data is from ChemCAD and simply refers to the method outlined in Chapter 13.

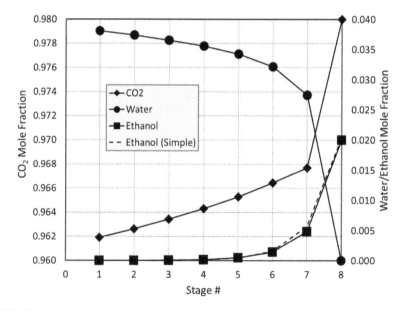

FIGURE 15.5
Vapor composition profiles for the seven-stage absorption column in Example 15.3. Data is from ChemCAD and simply refers to the method outlined in Chapter 13.

TABLE 15.3

Molar Flow Rates (mol/s) of Each Component in the Seven-Stage Absorber

		Liquid			Gas		
Stage	T	W	E	C	W	E	C
0	30.0	200.00	0.00	0.00			
1	29.8	199.93	0.00	0.11	3.87	0.00	97.87
2	29.4	199.84	0.00	0.11	3.80	0.00	97.98
3	29.1	199.74	0.01	0.11	3.71	0.01	97.98
4	28.6	199.60	0.05	0.11	3.61	0.02	97.98
5	28.0	199.38	0.14	0.12	3.48	0.05	97.98
6	26.9	198.91	0.49	0.12	3.26	0.15	97.98
7	24.4	196.13	2.00	0.13	2.78	0.49	97.99
8	30.0				0.00	2.00	98.00

the same resulted in temperatures dropping more through the column from 29.3°C to 19.0°C. Carbon dioxide has the same effect but there is so much less dissolved.

The amount of ethanol in the gas decreases sharply at the bottom of the column where the gas first contacts the liquid. By the middle of the column it is almost all absorbed. For this reason, the ethanol in the liquid doesn't start to increase until the middle of the column because the gas it contacts at the top simply doesn't have any ethanol left. In fact this seven-stage column removes 99.98% of the ethanol. Compare this to 99.2% for a four-stage, 98.1% for a three-stage, 94.1% for a two-stage, and 78.4% for a one-stage absorber (all ChemCAD results).

The ethanol profile in both the liquid and gas is also plotted in Figures 15.4 and 15.5 for the simplified analysis in Chapters 13 and 14. Note that it is almost identical to the rigorous analysis performed by ChemCAD. Of course the temperature, water, and carbon dioxide profiles are completely different because of the assumptions. At least for this case, it did not much affect the performance of the column with respect to ethanol removal.

Meanwhile, the water in the gas increases sharply at the bottom of the column where it first contacts the liquid and steadily increases toward the top, assisted by the increase in temperature that increases its vapor pressure.

The carbon dioxide appears to saturate the liquid at the bottom of the column where it first contacts the gas and decreases slightly going up the column because the increase in temperature reduces its solubility.

15.3 Summary

The simplifying assumptions in our prior absorption and stripping analysis in Chapters 13 and 14 were that a column was isothermal and that only the solute (in our example, the ethanol) transferred between phases. Clearly, this is an obvious simplification because we know that water evaporates, carbon dioxide will carbonate water, and there are cooling and heating effects involved in evaporation and dissolution/condensation. The more rigorous analysis in ChemCAD and our spreadsheet simulation for one- and two-stage absorbers show that there are significant differences, but not for ethanol removal performance. A chemical engineer will appreciate the tradeoffs between the ease of calculation with simplified methods and the more accurate results with rigorous calculations.

16

Column Design

Tray and packed columns are used for both distillation and absorption/ stripping processes. A qualitative description was given in Chapter 1. Once you've calculated how many equilibrium stages you need, why not just go down to your local separation column store and buy one? Well, there are still lots of design choices to make.

Should you use a tray column? What type of tray – sieve, bubble cap, or valve tray? What type of bubble cap or valve? (These days, few new columns would use bubble caps.) How far apart should they be spaced? How big are the holes, how many are there, how are they arranged on the tray? What kind of downcomer is used, how many per tray should there be, how much area do they occupy? What type and size of weir? And, very important, how many trays should you actually use?

Maybe you decide to use a packed column. Should you use random, dumped packing like Pall rings, Intalox saddles, or any one of a million (OK, maybe I'm exaggerating!) different patented designs? Should they be made from plastic, ceramic, metal, or glass? How much packing should be used? Maybe structured packing would be better, but there again, what type should you buy? What kind of internals should be used to distribute the liquid, hold up the packing at the bottom, and hold it down at the top?

Then there are issues with the column itself. How big a diameter is needed? How much room is needed for disengaging liquid from vapor at the top and for liquid surge capacity at the bottom. For distillation, how are the condenser and reboiler designed?

Fortunately, chemical processes are designed by teams of engineers. If you are a new chemical engineer who simply uses separation processes in the plant, the information in this book may be all you need to start – plus an appreciation for all the other issues raised. If you will specialize in separation processes you will want to read more comprehensive books, take advanced and graduate level courses, attend relevant workshops and professional meetings, and apprentice to company experts. Don't hesitate to rely on the experience of others by reading the technical literature, hiring consultants, discussing with vendors, and subscribing to services such as Fractionation Research Inc. Other engineers are needed to design the control system for safe and optimal operation, select materials for the column and internals that are compatible with the chemicals, design the structure so this 200-ft tower doesn't tip over (!), and the list goes on.

In this chapter, I will give you just a taste of separation column design. The two main issues will be "How tall should we make it?" and "How big should we make it?" To put it simply, the height depends on the number of stages and the diameter depends on flow rates.

Keep in mind that many methods have been developed to predict efficiency and diameter for both tray and packed columns. I have selected some well-known methods, but your future employer might very well use different methods, including mass transfer-based methods that I don't include.

16.1 How Tall Should a Column Be?

You are now pretty awesome at calculating the number of equilibrium stages required to accomplish a separation. For that to be the actual number of trays, they would need to be 100% efficient, but that is rarely the case. For a packed column with no discrete sections, what amount of packing even constitutes a stage? We need methods to estimate the efficiency of trays and packing. Efficiency is best estimated using actual data from similar systems. Absent that data, we have empirical correlations and rules of thumb. There are theoretical models that use mass transfer concepts, but they are complex and not particularly more accurate than empirical models. I will refer you to more comprehensive books for that discussion (Seader et al., 2015, Wankat, 2017).

Efficiency depends strongly on liquid viscosity, so distillation with higher temperatures tends to have higher efficiencies than absorption/stripping with lower temperatures. For example, a hydrocarbon at its boiling point in a distillation column might have a viscosity of 0.2 cP while water at room temperature in an absorber has a viscosity of 1 cP. The most common correlation for tray columns predicts an efficiency of around 70% for the distillation and 50% for the absorption. Imagine that you might need to double the number of trays you calculated by the Lewis or McCabe–Thiele analysis!

The height of a tray column depends on the actual number of trays, their spacing, and extra room for disengaging phases at the top and liquid surge capacity at the bottom. If typical spacing is 2 ft, with 4 ft at the top and 6 ft at the bottom, a 40-tray column will be about 90 ft tall. The height of a packed column depends on the height of packing required to achieve the specified separation, extra room at the top and bottom, and various internal structures that are hard to generalize.

16.2 How Tall Should a Tray Column Be?

Once the number of equilibrium stages has been determined, the actual number can be estimated using an efficiency. We have several ways to estimate the efficiency.

16.2.1 Overall Column Efficiency

We can define an overall column efficiency, E_0, to be

$$E_0 \equiv n_T/n_A \qquad (16.1)$$

where n_T is the theoretical number of stages (what you have been calculating) and n_A is the actual number of stages required to do the job. E_0 can be estimated using industrial experience with similar systems, empirical models from experimental data, or scale-up from lab and pilot plant operation.

16.2.2 Murphree Efficiency

One of the most common methods to determine the actual number of trays is to use the Murphree vapor efficiency (Murphree, 1925). For a particular stage

$$E_{MV} = \frac{y_{out} - y_{in}}{y^*_{out} - y_{in}} \qquad (16.2)$$

where y^*_{out} is the vapor composition that would be in equilibrium with the leaving liquid stream with composition, x_{out}, but the actual y_{out} hasn't quite reached that value. (A similar form, E_{ML}, is used for the liquid phase instead). In a McCabe–Thiele diagram that is equivalent to stepping from the operating line at (x_N, y_{N+1}) to the equilibrium line at (x_N, y^*_N) but only going part way in the vertical direction to (x_N, y_N) as illustrated in Figure 16.1.

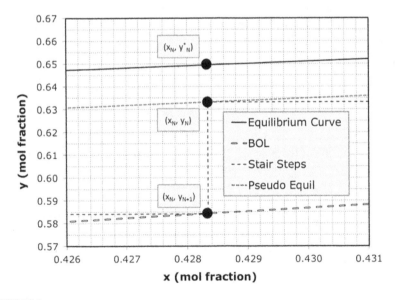

FIGURE 16.1
Detail of a McCabe–Thiele distillation design. Distance from (x_N, y_{N+1}) to (x_N, y_N) divided by distance from (x_N, y_{N+1}) to (x_N, y^*_N) is the Murphree vapor efficiency, E_{MV}.

A pseudo-equilibrium curve can be drawn with every vapor composition located at that percentage difference between the operating and equilibrium curve. Then, the stages are simply stepped off as usual. For our purposes, E_{MV} will be simply given, although it can be estimated using mass transfer concepts or backed out of experimental data.

EXAMPLE 16.1   Using the Murphree Vapor Efficiency

A total of 100 mol/s of a saturated liquid mixture of 45 mol% benzene and 55 mol% toluene is fed to the optimal tray of a distillation column operating at 1 atm and equipped with a total condenser (TC) and partial reboiler (PR). About 40 mol/s of a saturated liquid distillate with 0.95 mol B/mol is produced using a reflux ratio R = 2.0. The trays have a Murphree vapor efficiency of 0.75. How many trays are required and what is the optimal feed tray?

a. Draw a labeled process flow diagram and
b. Conduct a degree of freedom analysis.
    These steps will be the same as in Example 7.1. The problem can be solved graphically with the McCabe–Thiele method. Instead of stepping from the operating line to the equilibrium line, we only step 75% of the vertical distance.
c. Write the equations and solve.
    An overall process mole balance shows that B = 60 mol/s and $x_B = 0.117$ mol B/mol. The top operating line (TOL) can be drawn from the distillate composition $x_D = 0.95$ to the y-intercept at

$$y_{int} = \frac{x_D}{1+R} = \frac{0.95}{1+2} = 0.317$$

The q line is a vertical line (because it is a saturated liquid) at the feed composition z = 0.45. The bottom operating line (BOL) can then be drawn from x_B to the intersection of the q line and TOL. These three lines are drawn in Figure 16.2. Also shown are the stair steps that will now be explained. Why not follow along by printing the chart in Excel "Murphree/Op Lines" and stepping off the stages with me?
    Starting at the bottom with $x_B = 0.117$, the boilup vapor has a composition in equilibrium with the bottoms. Usually we consider the reboiler to be perfectly efficient so the vapor composition lies on the equilibrium line at $(x_B, y_R) = (0.117, 0.240)$. The liquid entering the reboiler from the bottom tray, L_N, is a passing stream to V_R and will lie on the BOL. It is on a horizontal line from (x_B, y_R) to (x_N, y_R) so $(x_N, y_R) = (0.199, 0.240)$.

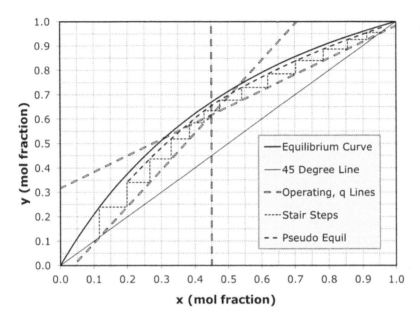

FIGURE 16.2
McCabe–Thiele diagram for Example 16.1 using $E_{MV} = 75\%$.

In previous chapters, we found the vapor leaving Stage N by drawing a vertical line from $(x_N, y_R) = (0.199, 0.240)$ to the equilibrium line at $(x_N, y_N) = (0.199, 0.374)$ because the two streams were considered to be in equilibrium. But because these trays are only 75% efficient we only draw the line 75% of the vertical distance to $(x_N, y_N) = (0.199, 0.340)$.

Would you like to see a few more steps? Streams L_{N-1} and V_N are passing streams with compositions on the BOL. So draw a horizontal line from $(x_N, y_N) = (0.199, 0.340)$ to $(x_{N-1}, y_N) = (0.266, 0.340)$. Streams L_{N-1} and V_{N-1} are leaving streams and *if they reached equilibrium* they would have the composition $(x_{N-1}, y_{N-1}) = (0.266, 0.468)$. So draw a vertical line from $(x_{N-1}, y_N) = (0.266, 0.340)$ to 75% of the distance to the equilibrium curve at $(x_{N-1}, y_{N-1}) = (0.266, 0.436)$.

Continue stepping off stages, crossing from the BOL to the TOL at the appropriate stage, until the distillate composition, $x_D = 0.95$, is reached. The "pseudo-equilibrium" curve is drawn for your convenience with all vapor composition values at 75% of the vertical distance between the operating lines and the equilibrium curve. The completed McCabe–Thiele diagram is shown in Figure 16.2. If the trays were 100% efficient, only nine trays would be necessary. With $E_{MV} = 75\%$, 12 trays are required. The optimal feed is to Tray 8. In both cases, a reboiler is also present and acts like an equilibrium stage.

d. Explore the problem.
   If the overall efficiency was given as 75%, then the number of required trays would be

$$n_A = n_T/E_0 = 9/0.75 = 12$$

This is the same number found using the Murphree vapor efficiency. Try lots of process conditions using Excel "Murphree/E_{MV}" to see if this is generally true (only under certain conditions).

e. Be awesome!
   Study the construction of Excel "Murphree/E_{MV}" and see if you can construct a simulation using the Murphree liquid efficiency in Excel "Murphree/E_{ML} DIY." In this case the horizontal distances will be shortened. Start at the top since the first step is a horizontal move.

16.2.3 O'Connell Correlation

An empirical model that has been the industry standard for the last 7 decades was provided by O'Connell (1946) who correlated the overall efficiency for bubble cap and some sieve tray columns with the relative volatility, α (unitless), and the liquid viscosity, μ_L (cP). Trays are less efficient at a large liquid viscosity since diffusion is slower, decreasing mass transfer rates. It is also less efficient with a large relative volatility/solubility because of the increased amount of material that must be transferred. The data was fitted with the well-known O'Connell correlation.

$$E_0 = 0.503(\alpha\mu)^{-0.226} \tag{16.3}$$

Duss and Taylor (2018) noticed that theoretical models would not place such emphasis on the relative volatility and modified the equation for a better fit of the data.

$$E_0 = 0.503(\mu)^{-0.226}(\alpha)^{-0.08} \tag{16.4}$$

Typically, you will calculate the efficiency at the average temperature and pressure of the column. Notice that flow rates are not part of this correlation – the efficiency is relatively constant over a wide range of flow rates, roughly 60%–80% of flooding velocities. Bubble cap trays are less efficient than sieve or valve trays so the resulting efficiency is a conservative value.

EXAMPLE 16.2 Using the O'Connell Correlation

Estimate the efficiency of the column presented in Example 16.1. A total of 100 mol/s of a saturated liquid feed of 45 mol% benzene and balance toluene was fed to the optimum stage of a ten-stage column equipped with a TC and PR. About 40 mol/s of distillate was produced with a reflux ratio $R = 2$.

a. Draw a labeled process flow diagram and
b. Conduct a degree of freedom analysis.
   These steps were completed in Example 7.1.
c. Write the equations and solve.
   A McCabe–Thiele analysis (Excel "Lewis/10-Tray") showed that the product compositions and temperatures are $x_D = 0.964$ mol B/mol at 80.8°C and $x_B = 0.107$ mol B/mol at 105.8°C for an average temperature of 93.3°C. At this temperature $\mu_B = 0.291$ cP and $\mu_T = 0.299$ cP. The average viscosity and relative volatility at the feed composition is

$$\mu_{mix} = \exp\{x_B \ln\mu_B + x_T \ln\mu_T\}$$

$$= \exp\{(0.45)(\ln 0.291) + (0.55)(\ln 0.299)\} = 0.296 \text{ cP}$$

$$\alpha = (y_B/x_B)/(y_T/x_T) = (0.67/0.45)/(0.33/0.55) = 2.5$$

$$E_0 \sim 0.503\mu^{-0.226}\alpha^{-0.08} = 0.71$$

In this case, the actual number of trays would be

$$n_A = n_T/E_0 = 10/0.71 = 14.1 \sim 14 \text{ trays}$$

d. Explore the problem.
   Calculate the efficiency at both the top and bottom of the column. Using the most conservative value, how many trays are required for this separation?

e. Be awesome!
   Duss and Taylor (2018) go on to suggest using the stripping factor ($\lambda = mG/L$) instead of the relative volatility based on theoretical considerations (m is the slope of the equilibrium curve and G and L are gas and liquid molar flow rates). In this case

$$E_0 = 0.503\mu^{-0.226}\sigma^{-0.08} \tag{16.5}$$

where $\sigma = \lambda$ if $\lambda > 1$ or $\sigma = 1/\lambda$ if $\lambda < 1$. Estimate the efficiency using Equation 16.5 and compare to the previous estimates you made with Equation 16.4.

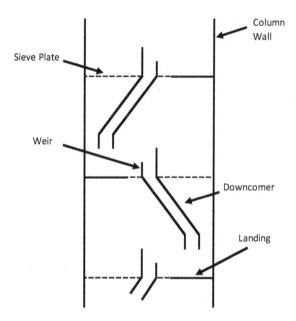

FIGURE 16.3
Oldershaw column with sieve plate and tubular downcomer.

16.2.4 Estimating Efficiency Using Laboratory Data

Efficiency can also be estimated from laboratory data, often using an Oldershaw column, shown in Figure 16.3. These columns are often made with glass and contain sieve trays with small tubular downcomers.

One way to use this laboratory scale column is to measure the separation at total reflux (no feed or products – just the initial inventory) and then use the McCabe–Thiele analysis to determine the theoretical number of trays required. The efficiency is calculated by dividing the number of theoretical trays by the number actually in the Oldershaw column.

A large column will usually have a greater efficiency than a laboratory Oldershaw column so simply using the Oldershaw column's efficiency will result in a conservative design.

EXAMPLE 16.3   Using an Oldershaw Column

The benzene/toluene mixture in the previous examples is distilled in an Oldershaw column operating at total reflux. The column features eight sieve trays, a TC, and a total reboiler. A distillate composition of 0.95 mol B/mol and a bottoms composition of 0.08 mol B/mol are produced. Estimate the efficiency of a valve tray column producing the specified amounts assuming that it operates at 80% of flooding. This

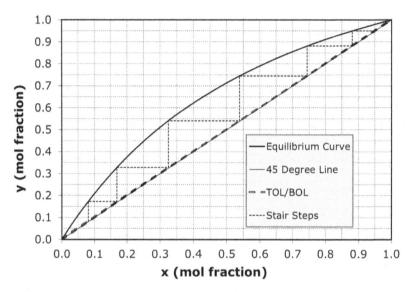

FIGURE 16.4
McCabe–Thiele diagram for an Oldershaw column operating at nearly total reflux.

type of problem will not require a systematic analysis. The theoretical number of trays at total reflux can be found from a McCabe–Thiele analysis. The result is shown in Figure 16.4.

Five equilibrium trays and a total reboiler are required using a reflux ratio of 100 (effectively total reflux). The efficiency is $E_0 = 5/8 = 63\%$. We can use this value to make a conservative design of the commercial column.

16.3 How Tall Should a Packed Column Be?

The methods for tray columns won't work for packed columns because there are no discrete sections that come close to equilibrium. Instead, we can talk about the HETP (height equivalent to a theoretical plate) (Peters, 1922), that describes the packing height (not the total volume) needed to approximate one theoretical stage. It is best to estimate this based on experience with the same or similar systems, keeping in mind that it is particularly sensitive to liquid viscosity, the type and size of packing material used, and liquid distribution. Absent data, several rules of thumb have developed over the years. Theoretical models based on mass transfer concepts are also available but are complex and not particularly accurate. Again, I will refer you to more comprehensive books for that discussion (Seader et al., 2015, Wankat, 2017).

16.3.1 The HETP Method

The HETP is defined as

$$\text{HETP} \equiv h_P/n_T \tag{16.6}$$

where h_P is the height of the packing and n_T is the number of theoretical stages as determined by a conventional McCabe–Thiele design.

The HETP can vary from a few inches to several feet and tends to be larger for larger packing. Some useful heuristics described by Kister (1992) relate HETP to column (D_T) or packing (D_P) size:

1. HETP (ft) = 1.5 D_P (in) for random packing and low-viscosity liquids.
2. HETP (ft) = 1.5 D_P (in)+0.5 for vacuum operation.
3. HETP (ft) = 5–6 for absorption with viscous liquids.
4. HETP (ft) = D_T (ft) if column is less than 2 ft diameter.
5. HETP (ft) = 1 if column is less than 1 ft diameter.
6. HETP (m) = $100/a_P$+0.10 for structured packing and low-viscosity liquids. a_P is the surface to volume ratio in m^2/m^3. Rules #1 and #2 have been updated for modern packings to use the surface area per volume because of the difficulty in defining a diameter for the complex shapes. Also, high surface tension liquids like water may require much larger (twice!) HEPT values due to underwetting of the packing.

EXAMPLE 16.4   Using the HETP

The absorption column in Example 14.1 used water to absorb 2 mol% ethanol from a gas stream of carbon dioxide with a total gas flow rate of 100 mol/s. About 95% of the ethanol was removed with seven equilibrium stages. It is suggested to use an existing 2 ft diameter column filled with 20 ft of 2.5 in Pall rings. Is this reasonable?

A rule of thumb is to use a packing size such that the column to packing diameter ratio is above 8 to avoid wall channeling and below 40 to avoid poor distribution of liquid. For 2.5 in Pall rings, the ratio is 9.6 so these are acceptably sized. The HETP would be approximately 5–6 ft for absorption with a viscous liquid. Let's use the smaller value of 5 ft since water is not particularly viscous. The packed height is (5 ft)(7) = 35 ft. A much taller column must be used to accomplish this separation. Whether a 35-ft tall and 2-ft diameter column filled with 2.5 in Pall rings can handle these flow rates is a question we have yet to answer.

16.4 How Big Should a Column Be?

The next "big" question (see what I did there?) is how large the column diameter should be. Both tray and packing towers are treated similarly except for using different correlations. The columns are sized so that they can handle the expected flow rates without flooding. This occurs when the flow rates are so large that the two phases cannot push past each other and start to back up. There are many non-ideal flow patterns that decrease efficiency, such as weeping at low gas flow rates that allows liquid to leak through the tray holes. The columns are designed for high gas flow rates, but not so high that entrainment carries liquid up the column with the gas.

Both methods presented here correlate the flooding velocity with a flow factor developed by Sherwood et al. (1938), which is a ratio of the two flow's kinetic energy.

$$F_{LG} \equiv \frac{LM_L}{GM_G} \sqrt{\frac{\rho_G}{\rho_L}} \tag{16.7}$$

L and G are liquid and gas molar flow rates, respectively, M is the molecular weight, and ρ is the density. These are called generalized pressure drop correlations (GPDCs) that have evolved significantly over the years.

16.5 How Big Should a Tray Column Be?

For a given liquid flow rate, we can estimate the gas flow rate that causes flooding by entrainment using the method developed by Fair and Matthews (1958). The flooding velocity, u_f (ft/s), is found using

$$u_f = C_{SB} \left(\frac{\sigma}{20} \right)^{0.2} \sqrt{\frac{\rho_L - \rho_G}{\rho_G}} \tag{16.8}$$

where σ is the surface tension of the liquid (dyne/cm), ρ is the density of the liquid or gas, and C_{SB} is a capacity factor that was correlated by Souder and Brown (1934) with the flow factor. The correlation by Fair and Matthews (1958) for sieve trays with various spacing is shown in Figure 16.5.

If we know the flow rates, tray spacing, and physical properties like density and viscosity, the flooding velocity can be found. The operating velocity, u_{OP}, will be best at some fraction, f, of the flooding velocity – say 75%.

$$u_{OP} = f u_f$$

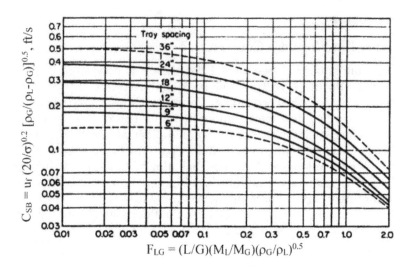

FIGURE 16.5
Capacity factor for flooding of sieve trays from Fair, J. R., and R. L. Matthews, "Better Estimate of Entrainment from Bubble Caps," Pet. Refiner, 37 (4), 153 (1958).

u_{OP} can be related to the area of the tray and therefore diameter of the column.

$$u_{OP} = \frac{GM_G}{\rho_G A} \qquad (16.9)$$

M_G is the molecular weight and A is the net area through which gas can flow (active area). Gas cannot flow through the downcomer and the downcomer landing usually has no holes. The correlation already accounts for the space between holes, assuming the hole area is at least 10% of the active tray area and can be corrected if less than that. The downcomer area usually occupies something like 10% of the area and is corrected using a parameter $\delta \sim 0.9$. So, the tower diameter is

$$D_T = \sqrt{\frac{4GM_G}{\pi \delta \rho_G u_{OP}}} \qquad (16.10)$$

As with efficiency, this is best calculated at several locations in the tower and the most conservative value chosen.

EXAMPLE 16.5   Calculating Tray Column Diameter Using the Fair Correlation

Estimate the column diameter for a sieve tray column with 2 ft tray spacing described in Example 16.1 using conditions at the top of the column.

At the top of the column we have effectively pure benzene at ~81°C with $L = 80$ mol/s and $V = 120$ mol/s. At these conditions, the liquid density $\rho_L = 813\,\text{g/L}$ and the surface tension $\sigma = 21.1$ dyne/cm. Using the ideal gas law, the gas density is

$$\rho_v = \frac{pMW}{RT} = \frac{1\,\cancel{\text{atm}}}{} \left|\frac{78.11\,\text{g}}{\cancel{\text{mol}}}\right| \frac{\cancel{\text{mol}}\,\cancel{K}}{0.08206\,\text{L}\,\cancel{\text{atm}}} \left|\frac{}{354\,\cancel{K}}\right| = 2.69\frac{\text{g}}{\text{L}}$$

The flow factor is

$$F_{LG} = \left(\frac{L}{G}\right)\left(\frac{M_L}{M_G}\right)\left(\frac{\rho_G}{\rho_L}\right)^{0.5} = \frac{80}{120}\left(\frac{2.69}{813}\right)^{0.5} = 0.038$$

Since we assume pure benzene, the molecular weights cancel and the molar ratio is the same as the mass ratio. From Figure 16.5, the capacity factor is $C_{SB} = \sim 0.38$.

Solving for the flooding velocity yields

$$u_f = C_{SB}\left(\frac{\sigma}{20}\right)^{0.2}\left(\frac{\rho_L - \rho_G}{\rho_G}\right)^{0.5} = 0.38\left(\frac{21.1}{20}\right)^{0.2}\left(\frac{813 - 2.69}{2.69}\right)^{0.5} = 6.7\text{ ft/s}$$

If we want to operate at 75% of flooding velocity, the operating velocity is

$$u_{op} = (0.75)(6.7) = 5.0\text{ ft/s}$$

and if the downcomer correction is 90%, the diameter is

$$D_T = \sqrt{\frac{4GM_G}{\pi\delta\rho_G u_{OP}}}$$

$$= \left(\frac{4}{}\left|\frac{120\,\cancel{\text{mol}}}{\cancel{\text{s}}}\right|\frac{78.11\,\cancel{\text{g}}}{\cancel{\text{mol}}}\left|\frac{}{3.1416}\right|\frac{}{0.9}\left|\frac{\cancel{L}}{2.69\,\cancel{\text{g}}}\right|\frac{\cancel{\text{s}}}{5\,\text{ft}}\left|\frac{1\,\text{ft}^{3\cdot2}}{28.32\,\cancel{L}}\right|\right)^{0.5}$$

$$= 5.9 = 6\text{ ft}$$

The diameter should be rounded up to the next highest half of a foot.

d. Explore the problem.

Calculate the diameter at the bottom of the column assuming the molar flow rates are $L = 180$ mol/s, $G = 120$ mol/s, the temperature is $T = 105°C$, and the composition is 90 mol% toluene with balance benzene. At these conditions, the liquid density $\rho_L = 813\,g/L$ and the surface tension $\sigma = 18.0\,dyne/cm$.

e. Be awesome

Calculate the flooding velocity of the absorption column described in Example 16.4 if it was equipped with sieve trays spaced 18 in apart. Can the 2 ft diameter column handle the specified flows?

16.6 How Big Should a Packed Column Be?

The diameter of a packed column is also based on the gas velocity that would cause flooding. It can be designed to operate at a percentage of that velocity or at a specified pressure drop (Kister et al., 2007). We'll use the generalized pressure drop correlation, originally developed by Sherwood et al. (1938) to show flooding conditions, and modified several times to include pressure drop, leading to a current version by Strigle (1987). The correlation is shown in Figure 16.6. The superficial gas velocity, u_S (ft/s), is determined from the ordinate and is a function of the same flow parameter on the abscissa used for flooding in sieve tray columns.

$$F_{LG} = \frac{LM_L}{GM_G}\left(\frac{\rho_G}{\rho_L}\right)^{0.5}$$

(16.11)

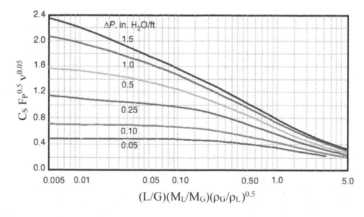

FIGURE 16.6
GPDC for random packings from Strigle, R. F. Jr., "Packed Tower Design and Applications," 2nd ed., Gulf Publishing, Houston, TX (1994).

L and G are liquid and gas molar flow rates, M is the molecular weight, and ρ is the density. Since these are all ratios, values with any consistent units are appropriate.

The ordinate is called the capacity factor and is

$$CP = C_S F_P^{0.5} v^{0.05} \tag{16.12}$$

F_P is the packing factor (ft^{-1}), which is characteristic of the type, size, and material of the packing. It is the surface area per unit volume, so more properly the units are (ft^2/ft^3). Table 16.1 is very brief excerpt from Table 14.13 in Green and Southard (2019) that lists characteristics of many random packing materials. (Be careful – F_P is given in unit m^{-1}.) v is the liquid kinematic viscosity (cSt) that is obtained by dividing the dynamic viscosity (cP) by the liquid density (g/cm^3). C_S is the superficial gas velocity, u_S (ft/s), corrected for the densities.

$$C_S = u_S \left[\rho_G / (\rho_L - \rho_G) \right]^{0.5} \tag{16.13}$$

Once u_S has been determined for a specified pressure drop (or vice versa), the area is calculated by

$$A \left(ft^2 \right) = GM_G / (u_S \rho_G) \tag{16.14}$$

and the tower diameter by

$$D_T (ft) = (4A/\pi)^{0.5} \tag{16.15}$$

The flooding line was originally on the GPDC graph above the curve for $\Delta P = 1.5$ in H$_2$O/ft but was omitted from the Strigle chart. However, the pressure drop at flooding ΔP_f, is easily calculated according to Kister (1991).

$$\Delta P_f = 0.12 F_P^{0.7} \tag{16.16}$$

TABLE 16.1

Characteristics of Random Packings

Type/Material	Size (mm)	Density (kg/m³)	Area (m²/m³)	Voids (%)	F_P (m⁻¹)
Pall rings/plastic	15	95	350	87	320
	25	71	206	90	180
	40	70	131	91	131
	50	60	102	92	85
	90	43	85	95	56
Intalox/metal	25	224	207	97	134
	40	153	151	97	79
	50	166	98	98	59
	70	141	60	98	39

Excerpted from Green, D. W. and M. Z. Southard, *Perry's Chemical Engineers Handbook*, 9th ed., McGraw-Hill, 2019.

EXAMPLE 16.6   Estimating Absorption Column Diameter Using Strigle's GPDC Chart

A total of 100 mol/s of gas containing 2.0 mol% ethanol (E) and balance carbon dioxide (C) is contacted with pure water (W) in a seven-stage absorber packed with 90 mm plastic Pall rings and operating at 30°C and 825 mm Hg. About 95 mol% of the ethanol is absorbed. Flow rates and compositions were determined in Example 13.3. Estimate the required column diameter. Note that the number of stages (i.e., height of packing) does not enter into this calculation.

    a. Draw a labeled process flow diagram,

    b. Conduct a degree of freedom analysis, and

    c. Write the equations and solve.

        These steps were done in Example 13.3 so we will just use those results. The flow rates and compositions at the ends of the column are shown in Figure 16.7. Flooding is more likely at the end of the column with greater flow rates. That is at the bottom with G = 100.0 mol/s and y = 0.02 mol E/mol, and L = 71.4 mol/s and x = 0.0266 mol E/mol.

        The values inputted to the correlation are:

$$MW_L = (0.0266 \text{ mol E/mol})(46.07 \text{ g E/mol E})$$

$$+ (0.9734 \text{ mol W/mol})(18.02 \text{ g W/mol W}) = 18.77 \text{ g/mol}$$

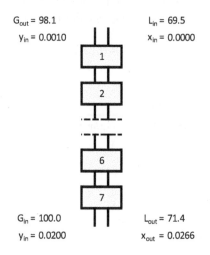

FIGURE 16.7
PFD of the seven-stage absorption column described in Example 16.4. G, L [=] mol/s, x, y [=] mol E/mol.

$MW_G = (0.0200 \text{ mol E/mol})(46.07 \text{ g E/mol E})$

$\quad + (0.9800 \text{ mol C/mol})(44.01 \text{ g C/mol C}) = 44.05 \text{ g/mol}$

$\rho_L = 1{,}000 \text{ g/L} = 62.4 \text{ lb/ft}^3$ (assume same as pure water)

$\rho_G = [(44.05 \text{ g/mol})(825 \text{ mm Hg})]/[(0.08206 \text{ L atm/mol K})$

$\quad (303 \text{ K})(760 \text{ mm Hg/atm})] = 1.92 \text{ g/L} = 0.120 \text{ lb/ft}^3$

Note that L, G, ρ_G, and ρ_L in the flow factor can be in SI or engineering units wherever the units cancel out.

$F_{LG} = (1.92/1{,}000)^{0.5}[(71.4)(18.77)/(100)(44.05)] = 0.013$

$F_P = 56 \text{ m}^{-1} = 17.1 \text{ ft}^{-1}$ (For 90 mm plastic Pall rings per Table 16.1.)

$\Delta P_f = (0.12)(17.1)^{0.7} = 0.88 \text{ in H}_2\text{O/ft packing}$

$CP =\sim 1.8$ (from Strigle chart)

$v = (1 \text{ cP})/(1 \text{ g/cm}^3) = 1 \text{ cSt}$

$C_S = 1.8/[(17.1^{0.5})(1^{0.05})] = 0.44 \text{ ft/s}$

$u_f = (0.44)[(1{,}000 - 1.92)/1.92]^{0.5} = 10.0 \text{ ft/s}$

$u_{OP} = (0.75)(10.0) = 7.5 \text{ ft/s}$ (operating at 75% of flooding velocity)

$A = [(100)(44.05)/454]/[(0.120)(7.5)]$

$\quad = 10.8 \text{ ft}^2$ (don't forget to convert gmol/s to lbmol/s)

$D_T = [4(10.8)/3.1416]^{0.5} = 3.7 \sim 4.0 \text{ ft} = 48 \text{ in}$

Double check the units on these calculations. However, the ordinate is not meant to be dimensionally consistent; just use the specified units for packing factor and viscosity. The tower diameter is increased to the next half foot increment. About 90 mm (~3.5 in) Pall rings were specified. The tower diameter to packing size ratio is

$D_T/D_P = 48/3.5 = 14$

The ratio should be at least 8 to avoid channeling along the wall where the particle bulk density is always lowest. These Pall rings will work.

d. Explore the problem.

Try using a smaller Pall ring and repeat these calculations. Discuss the results with your study partners.

e. Be awesome!

Try the Leva correlation (1992) as described in Seader, Henley, and Roper (2015) and the Eckert correlation (1970) as described in Wankat (2017). These are also GPDC models for column capacity estimations. How close are the diameters?

16.7 Summary

Designing a tray or packed column involves a huge number of decisions. The two factors that most influence cost are the diameter and height of the column. These can be estimated using prior experience with similar systems, laboratory and pilot plant data, empirical correlations, and rules of thumb. Several methods were demonstrated but this is only meant to give you a taste of column design. Methods date back as far as the 1920s and continue to be active areas of research, especially those based on mass transfer rates.

Part IV

Solvent Extraction

In Chapters 12–16, we looked at absorption and stripping, in which a second phase was added to cause separation of chemicals. For example, in stripping, a gas is added to a liquid and one or more components in the liquid diffuse into the gas. We used removal of ethanol from water by stripping with air as an example. Sometimes the process is complicated by air dissolving into the water and/or water evaporating into the air. Just like distillation, we first analyzed single equilibrium stages using a combination of mole balances, phase equilibrium relationships, and various process specifications. Then we added more stages and the accompanying additional – but not different – equations to analyze performance. Graphs of leaving and passing stream compositions suggested graphical techniques that we used to design systems.

We'll follow that same development for solvent extraction in which a second liquid, solvent, is added to a liquid feed consisting of a solute dissolved in a diluent. Some of the solute will diffuse into the solvent and reach equilibrium given enough time and contact area. The liquids should be immiscible – or at least partially immiscible – so that they can be separated. The product phase rich in diluent is called the raffinate. The product phase rich in solvent (and newly extracted solute) is called the extract. All three components, diluent, solute, and solvent can wind up in both phases. Although many chemicals can be involved in a solvent extraction process, we will limit this discussion to three. The process can be analyzed by the very same set of equations you know all too well by now – mole balances, phase equilibrium relationships, and various process specifications. Better separation can occur if the two liquid phases are contacted in countercurrent flow through a series of stages. The same equations are used to analyze the process, just more of them. Plotting the passing and leaving streams between stages will inspire a graphical technique that can be used to design a process.

No account will be given about stage efficiency and other design consider- ations. Separation processes require that the two phases separate by density differences. Whereas a liquid has a density of roughly 1,000 times a gas, two liquid phases have much closer densities. Take for instance a water-organic system with densities 1.0 versus 0.8 g/ml, respectively. Extra measures must be taken for efficient mixing and settling. Sometimes this is done in separate mixer and settler tanks. Other times it is done in tray or packed columns with extra mixing by hydraulic pulsing or mechanical rotation.

A challenge with solvent extraction is the complicated phase behavior. By definition, each liquid is non-ideal, so the phase equilibrium models are rather complex. Traditionally, this topic was taught using graphical methods only. We will start off like this, but I will show you a simple way to model the equilibrium for mathematical analysis.

Keeping with the usual progression of topics in this book, Chapter 17 "Single-Stage Solvent Extraction" starts off with graphical analysis of a single equilibrium stage. Then we analyze the stage using a completely empirical and a semi-empirical model of the phase equilibrium along with mole balances.

In Chapter 18 "Mathematical Analysis of Solvent Extraction Columns," we find that putting two or three stages together with the flows running countercurrent greatly increases the degree of separation. As with the other processes, plotting the passing stream compositions leads to a graphical design technique. The mathematical complexity of multi-stage processes starts to challenge Excel's Solver so it is best to solve for additional stages with a process simulator.

The graphical technique is demonstrated in Chapter 19 "Graphical Design of Solvent Extraction Columns." The algebraic basis is developed from mole balances around individual stages, resulting in a type of operating line. Combined with equilibrium tie lines, the Hunter–Nash method is demonstrated.

17

Single-Stage Solvent Extraction

A liquid mixture of chemicals can be separated by contacting it with a second liquid that is at least partially immiscible so that the two phases can be separated after some of the chemicals have transferred. In this chapter, we will analyze a single stage, much like extraction in a separatory funnel that you might have performed in a chemistry lab. Just like distillation, absorption, and stripping, the performance can be predicted with a combination of mole balances, phase equilibrium relationships, and other process specifications. However, about those phase equilibrium relationships ...

17.1 Liquid–Liquid Equilibrium

Vapor–liquid equilibrium (VLE) is described well for most systems by Raoult's law, or a modified Raoult's law if the system is not ideal. In the latter case, a model is used to calculate the liquid-phase activity coefficient. For the ethanol–water system, we used the Wilson equation. Describing liquid–liquid equilibrium (LLE) is more problematic because both liquids are non-ideal as evidenced by their forming two immiscible phases. The most common activity coefficient models for these systems are the semi-empirical NRTL (non-random, two liquid) and UNIQUAC (UNIversal QUAsiChemical). These are rather difficult to calculate. The NRTL equation will be presented in an optional section at the end of this chapter.

17.2 Graphical Analysis

Graphical analysis is the traditional way to introduce solvent extraction because the phase equilibrium behavior is rather complex. In fact, even commercial CAD software can fall quite short of predicting solvent extraction outcomes accurately. A ternary phase diagram can be drawn from experimental data for graphical analysis. Simple polynomial equations can be used to fit phase equilibrium data (instead of using the NRTL equations) for mathematical analysis.

Consider a mixture of water (diluent) and acetone (solute) that will be extracted with chloroform (solvent). The equilateral triangle phase diagram is shown in Figure 17.1. The apex of each side represents 100 mol% of the chemical. The sides represent just two of the three chemicals, for instance, the bottom of the triangle represents just the solvent and diluent with no solute. Any point within the triangle represents a combination of all three chemicals. The area bounded by the solid curve represents the two-phase region – any mixture within that will separate into two immiscible liquids. The dashed tie lines represent the compositions of those two liquids at equilibrium.

Figure 17.2 shows the same data presented as a right triangle phase diagram in which the acetone mol% is plotted against the water mol%. The chloroform mol% is just the amount needed to sum to 100 and can be thought of as represented by the 45° line. This graph is easier to generate in Excel and will be used here.

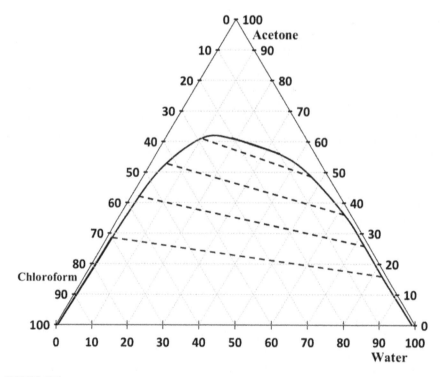

FIGURE 17.1
Equilateral triangle phase diagram for water (diluent), acetone (solute), and chloroform (solvent) at 25°C. Compositions are in mol%.

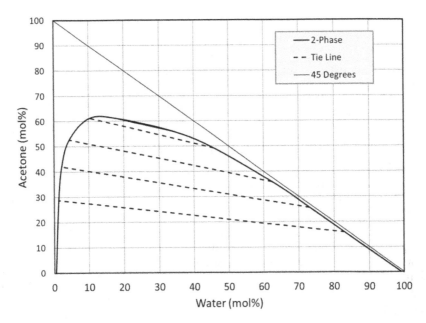

FIGURE 17.2
Right triangle phase diagram for water–acetone–chloroform at 25°C.

EXAMPLE 17.1 Generating an LLE Graph

J. M. Sorensen and W. Arlt (1980) published a large collection of experimental LLE data and extracted the NRTL and UNIQUAC parameters. For the water (diluent) – acetic acid (solute) – isopropyl acetate (solvent) system at 25°C, they provide the experimental tie lines shown in Table 17.1.

TABLE 17.1
Ternary Liquid–Liquid Equilibrium Data for Water (W)–Acetic Acid (A)–Isopropyl Acetate (I) at 25°C

Extract (mol frac)			Raffinate (mol frac)		
I	A	W	I	A	W
0.88690	0.00000	0.11310	0.00395	0.00000	0.99605
0.66806	0.10890	0.22304	0.00734	0.03881	0.95385
0.47325	0.18456	0.34219	0.01266	0.07648	0.91086
0.33804	0.21757	0.44439	0.02072	0.10854	0.87074
0.22825	0.22161	0.55014	0.03784	0.13861	0.82355
0.17964	0.21303	0.60732	0.05851	0.15815	0.78334

These data are plotted in Figure 17.3 as a right triangle phase diagram. Note that the compositions are in mole fraction rather than mole percent. The extract points (circles), raffinate points (squares), and the tie lines (dashed) are shown. The 45° line is shown for reference – any composition on this line has no isopropyl acetate. The points on each side are simply connected by a fourth-order polynomial equation (solid lines). The very bottom of the graph represents the solubility limits of water and isopropyl acetate without acetic acid. The last set of tie lines at the upper right hand corner is near the plait point where the extract and raffinate lines intercept.

More detailed information is shown in Figure 17.4. Line 1 represents all possible compositions when 20 mol% acetic acid in water (point F for feed) is mixed with pure isopropyl acetate (point S for solvent). If an equimolar mixture is prepared (point M for mixture), two phases will appear at equilibrium. If too little solvent (point B) or too much solvent (point C) is used, only one phase exists and separation is impossible. Also, if too much acetic acid is in the water (Line 2) no combination will be in the two-phase region.

The tie lines represent the compositions of the two phases at equilibrium. For any mixture that does not fall on these lines, the compositions can be determined by interpolation. The product line $\overline{ER}$

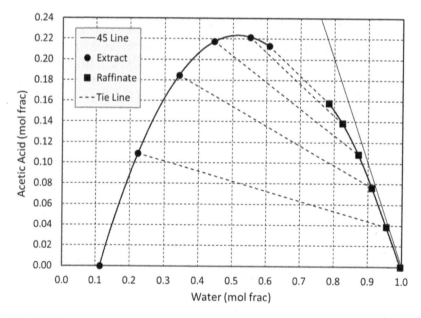

FIGURE 17.3
Ternary phase diagram for water–acetic acid–isopropyl acetate at 25°C.

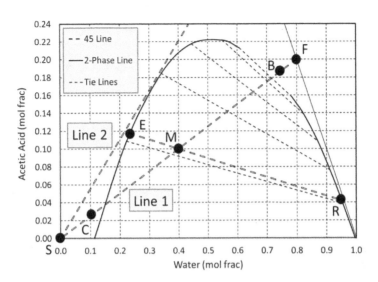

FIGURE 17.4
Ternary phase diagram for water–acetic acid–isopropyl acetate at 25°C showing additional features discussed in text.

represents the interpolated tie line for the equimolar mixture of feed and solvent described above. The extract (point E for extract) composition is roughly 23 mol% water, 12 mol% acetic acid, and 65 mol% isopropyl acetate. The raffinate (point R for raffinate) composition is roughly 95 mol% water, 4 mol% acetic acid, and 1 mol% isopropyl acetate.

EXAMPLE 17.2 Locating a Mixture

A total of 100 mol/s of feed consisting of 20 mol% acetic acid (A) and 80 mol% water (W) is mixed with 50 mol/s of pure isopropyl acetate solvent (S). Locate and label the compositions of the feed, solvent, and mixture on the phase diagram shown in Figure 17.5.

The feed has coordinates $(x_F, y_F) = (0.80, 0.20)$ and lies on the 45° line. The solvent has coordinates $(x_S, y_S) = (0.00, 0.00)$ and is located at the origin. The mixture contains

$$M = F + S = 100 + 50 = 150 \text{ mol/s}$$

where M, F, and S are the amount of mixture, feed, and solvent, respectively. The mixture composition (x_M, y_M) is

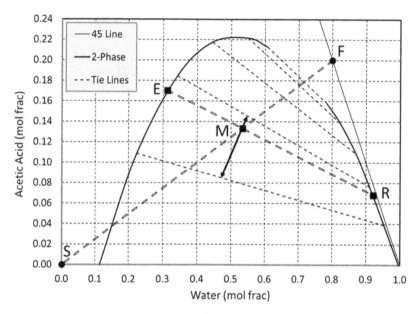

FIGURE 17.5

Ternary phase diagram for water–acetic acid–isopropyl acetate at 25°C as described in Example 17.2.

$$x_M = \frac{\left[\dfrac{0.80 \text{ mol W}}{\text{mol}}\right][100 \text{ mol/s}]}{[150 \text{ mol/s}]} = 0.533 \text{ mol W/mol}$$

$$y_M = \frac{\left[\dfrac{0.20 \text{ mol A}}{\text{mol}}\right][100 \text{ mol/s}]}{[150 \text{ mol/s}]} = 0.133 \text{ mol A/mol}$$

This is also plotted in Figure 17.5 as point M. Note that the mixture will always lie on the straight line drawn between the solvent and feed points. Believe me! What? You don't believe me?

The mole balances for water (x) and acetic acid (y) are

$$F x_F + S x_S = M x_M = (F + S) x_M = F x_M + S x_M \tag{17.1}$$

$$F y_F + S y_S = M y_M = (F + S) y_M = F y_M + S y_M \tag{17.2}$$

Collecting terms involving F and S from the left and right sides of these equations and then solving for F/S yields (work this out on paper!)

$$\frac{F}{S} = \frac{x_M - x_S}{x_F - x_M} = \frac{y_M - y_S}{y_F - y_M} \tag{17.3}$$

Rearranging into the form "rise/run" to find the slope yields

$$\frac{y_F - y_M}{x_F - x_M} = \frac{y_M - y_S}{x_M - x_S} \tag{17.4}$$

The slope of line segment $\overline{FM}$ is the same as the slope of line segment $\overline{MS}$ and the lines share point M. There is your proof that point M is always on the line between points F and S. You're welcome!

EXAMPLE 17.3   Using the Inverse Lever Rule

The greater the proportion of solvent, the closer will be the mix point to the origin and vice versa for the feed. Using the x-axis of the graph itself as a ruler, the segments $\overline{SF}$, $\overline{SM}$, and $\overline{MF}$ have lengths proportional to

$$\overline{SM} \propto x_M - x_S = 0.53 - 0.00 = 0.53$$

$$\overline{MF} \propto x_F - x_M = 0.80 - 0.53 = 0.27$$

$$\overline{SF} \propto x_F - x_S = 0.80 - 0.00 = 0.80$$

And so using the opposite segment (remember this is the *inverse* lever rule) the proportion of solvent is

$$S/M \sim 0.27/0.80 = 0.34$$

or 1/3 of the mixture and the proportion of feed is

$$F/M \sim 0.53/0.80 = 0.66$$

or 2/3 of the mixture. These are, of course, the actual proportions specified in the example.

We just solved for proportions while knowing the compositions. If you don't know the composition of the mix point but do know the proportions, draw a straight line between points S and F, and measure a distance between them proportional to the amount of each.

**EXAMPLE 17.4   Determining the Extract
and Raffinate Composition**

The tie lines represent equilibrium compositions. If the mix point was (0.500, 0.082) it would be on a tie line and it would be easy to find the extract and raffinate compositions. But if it is between tie lines, the compositions can be determined by interpolation. If you are in a hurry, it can be estimated by simply drawing a line with a slope somewhat intermediate between the adjacent tie lines. You can be a little more quantitative by drawing a line somewhat perpendicular to the adjacent tie lines and through the mix point as shown as the double arrow line in Figure 17.5. It will be the shortest line possible between the tie lines and going through the mix point. Measure the distance between the mix point and the tie lines on this line. Use that same proportion on the extract side and raffinate side to estimate the compositions. There are more exact methods, but graphical techniques are simply not that precise.

The mix point is roughly 80% of the distance between the lower and upper tie lines. This puts the extract and raffinate at points 80% between the tie line ends along their curves. They are indicated in Figure 17.5 as extract, point E (0.31, 0.17), and raffinate, point R (0.92, 0.69). Just like feed compositions S, M, and F, the product compositions E, M, and R all lie on the same line.

EXAMPLE 17.5   Find the Amount of Extract and Raffinate

The inverse lever rule can also be used to find the proportions of extract and raffinate. Connect points E and R in Figure 17.5. The line should pass through point M. Using the x-axis as a ruler, the amount of extract and raffinate are (remember the total flow of solvent and feed is 150 mol/s)

$$E = \left(\frac{\overline{MR}}{\overline{ER}}\right)M = [(0.92 - 0.53)/(0.92 - 0.31)]150 = 96 \text{ mol/s}$$

$$R = 150 - 96 = 54 \text{ mol/s}$$

You can also use simple mole balances.

$$\text{Water MB: } E(0.313) + R(0.921) = (100)(0.80)$$

$$\text{Overall MB: } E + R = 150$$

Solving for E and R will (of course) give the same amounts.

TRY THIS AT HOME 17.1   Using a Ruler to Measure Segment Lengths

In the above example, we used the x-axis as a ruler. Print the graph from Excel "Solvent Extraction/Ternary Diagram." Locate the solvent and feed compositions, label them S and F, and draw a line between them. Measure the length of the $\overline{SF}$ segment with an actual ruler – both in inches and centimeters. Measure 2/3 of this distance along $\overline{SF}$ and mark that point M. Is the composition (0.53, 0.13)? Does it matter what scale you use?

In Figure 17.6, I have used my handy cubit ruler. What the heck is a cubit? A cubit is an ancient measure of length defined as the distance from the elbow to the tip of the middle finger – roughly 18 in. So a centicubit (cc – don't confuse with the volumetric measure of cubic centimeters) is $18/100 = 0.18$ in. On the scale that I printed the graph, the line segment $\overline{SF}$ measures 25 cc. Measuring 2/3 of this distance, or 16.7 cc, from point S locates the correct mixture composition. Fun fact: you will not find the unit "centicubits" in any other modern textbook!

In Figure 17.7, line segment $\overline{ER}$ measures 17.0 cc and line segment $\overline{EM}$ measures 6.2 cc, so $R = (6.2/17.0)150 = 55$ mol. Keep in mind that a graphical solution will not be as precise as an analytical solution.

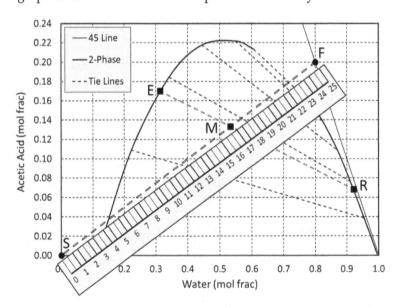

FIGURE 17.6
Using the inverse lever rule to determine the composition of a mixture from 100 mol of an 80 mol% water and 20 mol% acetic acid feed with 50 mol of isopropyl acetate. The ruler is in units centicubits (cc).

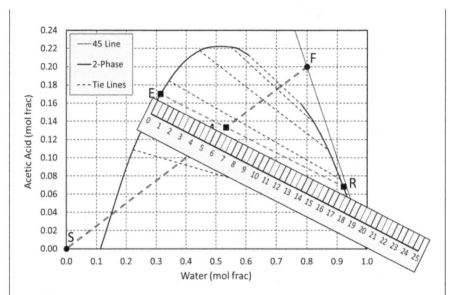

FIGURE 17.7
Using the inverse lever rule to determine the proportions of extract and raffinate from a mixture with composition $(x_M, y_M) = (0.53, 0.13)$. The ruler is in units centicubits (cc).

EXAMPLE 17.6   Single-Stage Graphical Analysis

A sample of contaminated water is partially purified by extraction with a solvent using a separatory funnel at 25°C. The amount treated is 20.0 mol containing 20.0 mol% acetic acid and the balance water. It is mixed with 20.0 mol of solvent containing 2.0 mol% acetic acid and the balance isopropyl acetate. After shaking for a minute, the funnel is placed back on a stand and the mixture separates into a top organic layer (extract) and a bottom aqueous layer (raffinate). Determine the amount and composition of the two layers.

a. Draw a labeled process flow diagram.
    This isn't a flow system, although we can imagine one by simply adding a time unit to the amounts. Still, it is useful to sketch the system as drawn in Figure 17.8. The water mole fraction is labeled x and the acetic acid mole fraction is labeled y. This is consistent with the labeling on a ternary phase diagram. The isopropyl acetate composition is not labeled since the mole fractions must add to 1.

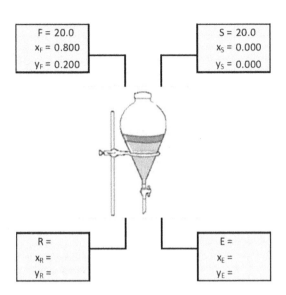

FIGURE 17.8
Single-stage laboratory solvent extraction for use in Example 17.6.

In practice, the denser aqueous layer will settle at the bottom and can be drained through the stopcock to a product container first. Then the organic layer can be drained into a separate container. There are many YouTube videos on laboratory solvent extraction that are valuable to view if you have never done this experiment.

b. Conduct a degree of freedom analysis.

Unknowns (6): E, x_E, y_E, R, x_R, y_R

Equations (6):

    3 mol balances (total, W, and A)

    3 equilibrium relationships (W, A, I)

There are zero DOF so the problem is solvable. The three phase equilibrium relationships are represented by the extract and raffinate curves along with the tie lines.

c. Write the equations and solve.

Use the graphical method as illustrated in Figure 17.9. Print a phase diagram from Excel "Solvent Extraction/Ternary Diagram" and follow along.

1. Locate the feed (F) and solvent (S) compositions.

2. Draw a line between these points.

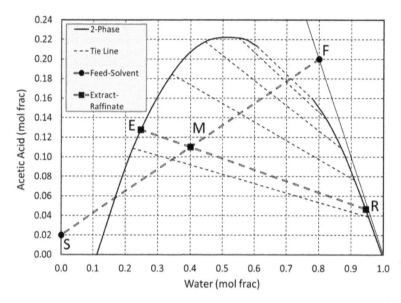

FIGURE 17.9
Ternary phase diagram for water–acetic acid–isopropyl acetate at 25°C with the graphical solution for Example 17.6.

3. Find the mix point (M) using a simple mole balance or the inverse lever rule with the x-axis as a ruler.

$$\overline{SF} \propto 0.8$$

$$\overline{SM}/\overline{SF} \propto 20/(20+20) = 0.50$$

$$\overline{SM} = (0.80)(0.50) = 0.40$$

4. Sketch a tie line through the mix point with an intermediate slope relative to the nearest tie lines.
5. Locate the extract (E) and raffinate (R) compositions. They are extract (0.25, 0.13) and raffinate (0.95, 0.05).
6. Use the inverse lever rule and/or simple mole balances to determine the amount of extract and raffinate.

$$E + R = 40$$

$$E(0.25) + R(0.95) = 20(0.80)$$

Solving the two equations yields E = 31.3 mol and R = 8.7 mol.

d. Explore the problem.

What is the minimum and maximum amount of this solvent you can add to this system and still have two phases? Find solvent-rich and feed-rich mixtures that lie outside the two-phase region.

e. Be awesome!

Can you even fit this much liquid in a 1-L separatory funnel? Look up the molecular weight and density of isopropyl acetate and assume that the liquids are pure solvent and water. If you are a third-year chemical engineering student and don't know the molecular weight and density of water, you might not be awesome yet!

TRY THIS AT HOME 17.2   Single-Stage Graphical Analysis

A total of 100 mol/s of a liquid containing 80 mol% water and the balance acetic acid is extracted in a single stage with 30 mol/s of pure isopropyl acetate. Determine the flow rate and composition of the extract and raffinate at equilibrium.

a. Draw a labeled process flow diagram.

A process flow diagram (PFD) is drawn in Figure 17.10. Fill in all known flow rates and compositions, leaving the

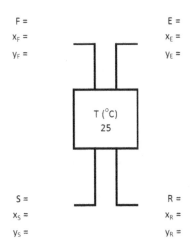

$F =$ $E =$
$x_F =$ $x_E =$
$y_F =$ $y_E =$

$T\ (^{\circ}C)$
25

$S =$ $R =$
$x_S =$ $x_R =$
$y_S =$ $y_R =$

FIGURE 17.10
PFD for single-stage solvent extraction described in Try This at Home: 17.2.

unknowns blank for now. As usual, the isopropyl acetate composition is not labeled and understood to be the balance.

b. Conduct a degree of freedom analysis.

Unknowns (6):

Equations (6):

You should find that there are zero DOF and the problem can be solved.

c. Write the equations and solve.

Use the graphical method with the extract, raffinate, and tie lines representing the three equilibrium relationships to determine compositions. Use the inverse lever rule and/or mole balances to determine the flow rates. The graphical solution is shown in Figure 17.11. Show that 72 mol/s

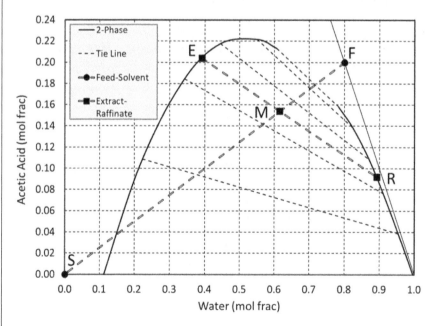

FIGURE 17.11
Graphical solution for the solvent extraction process described in Try This at Home 17.2.

of extract with composition (0.39, 0.20) and 58 mol/s of raffinate with composition (0.89, 0.09) are formed.

d. Explore the problem.

If you have been solving for flow rates using mole balances, rework all examples using the inverse lever rule or vice versa. If you have been working all examples using both methods – that is pretty awesome!

e. Be awesome!

Search for images of industrial solvent extraction processes – both photos and schematics. Study them carefully so that you understand process features. Add them to your separations photo album.

17.3 Mathematical Analysis

If we have algebraic equations that describe the phase equilibrium, a spreadsheet simulation of a solvent extraction process should be like those we have already developed previously for distillation, absorption, and stripping. Equations of state such as the NRTL model do exactly that. However, a simpler, completely empirical approach is to fit the experimental data with polynomial functions.

The experimental data in Table 17.1 were plotted in Figure 17.3. The extract and raffinate compositions were fitted to polynomial functions. A third-order polynomial works but a more precise fourth-order polynomial is used here and results in $R^2 = 1$.

$$y = ax^4 + bx^3 + cx^2 + dx + e \tag{17.6}$$

The variable y refers to the acetic acid mole fraction and the variable x refers to the water mole fraction. If a polynomial equation does not fit the data well, you can split the extract and/or raffinate curves into parts and fit with multiple equations.

These describe the raffinate and extract compositions but not the equilibrium relationship between them. For this we can calculate the slope of each tie line (including zero slope for no acetic acid) and fit it as a polynomial function of one of the compositions – here I chose water in the extract. So, the range of this function is from the acetic acid-free $x_E \sim 0.12$ to near the plait point $x_E \sim 0.60$. The tie line slope versus extract water mole fraction is plotted in Excel "Solvent Extraction/Tie Line Slope."

The constants for all three polynomial functions are shown in Table 17.2. Any extract and raffinate compositions at equilibrium must satisfy all three equations, instead of using NRTL phase equilibrium equations for each of the three components.

TABLE 17.2

Constants for Polynomial Functions Describing LLE for Water (W)–Acetic Acid
(A)–Isopropyl Acetate (I) at 25°C

	$y = ax^4 + bx^3 + cx^2 + dx + e$				
	a	b	c	d	e
Slope	−2.4926	2.3586	−1.2139	1.1292	−0.1152
Extract	0.1877	0.0679	−1.6418	1.5315	−0.1523
Raffinate	13.674	−44.847	53.288	−27.628	5.5099

EXAMPLE 17.7   Solvent Extraction Mathematical Analysis

A total of 100 mol/s of a liquid feed containing 80 mol% water and
20 mol% acetic acid is to be extracted with pure isopropyl acetate. The
process is carried out at 25°C and 1000 mm Hg. Determine the flow
rates and compositions of the raffinate and extract products.

a. Draw a labeled process flow diagram.
   The labeled PFD is shown in Figure 17.12. Note that the
   isopropyl acetate mole fraction is not labeled since it simply
   sums to 1 with the other mole fractions.

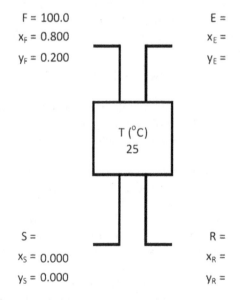

$$F = 100.0$$
$$x_F = 0.800$$
$$y_F = 0.200$$

$$E =$$
$$x_E =$$
$$y_E =$$

$$T\,(°C)$$
$$25$$

$$S =$$
$$x_S = 0.000$$
$$y_S = 0.000$$

$$R =$$
$$x_R =$$
$$y_R =$$

FIGURE 17.12
Process flow diagram for the single-stage solvent extractor described in Example 17.7.

b. Conduct a degree of freedom analysis.

Unknowns (7): S, E, x_E, y_E, R, x_R, y_R

Equations (6):

    3 mole balances (total, water, and acetic acid)

    3 phase equilibrium relationships (represented by the 3 polynomial equations).

There are 7 unknowns and 6 equations, leaving one DOF. In order to analyze this process, we need to specify either a solvent flow rate or an outcome. Let's use a process specification: 90 mol% of the acetic acid in the feed must be removed.

c. Write the equations and solve.

Overall mole balance: $F + S = E + R$

Water mole balance: $Fx_F = Ex_E + Rx_R$

Acetic acid mole balance: $Fy_F = Ey_E + Ry_R$

Extract: $y_E = a_E x_E^4 + b_E x_E^3 + c_E x_E^2 + d_E x_E + e_E$

Raffinate: $y_R = a_R x_R^4 + b_R x_R^3 + c_R x_R^2 + d_R x_R + e_R$

Slope: $(y_E - y_R)/(x_E - x_R) = a_S x_E^4 + b_S x_E^3 + c_S x_E^2 + d_S x_E + e_S$

Process specification: (90% removed): $Ey_E = 0.90 Fy_F$

These equations can be solved using Excel "Solvent Extraction/1-Stage." You should find that 97.8 mol/s of solvent is required. About 151.1 mol/s of extract (0.236, 0.119) and 46.7 mol/s of raffinate (0.950, 0.043) are formed. Fill in the rest of the unknowns in Figure 17.12. The graphical solution is shown in Figure 17.13.

d. Explore the problem.

    Make up at least five different problem statements with a variety of specifications. Print out the blank phase diagrams in Excel "Solvent Extraction/Ternary Graph" and solve the problems graphically. Make the appropriate modifications to the spreadsheet and solve them analytically. Note that the graph in Excel "Solvent Extraction/1-Stage Graph" automatically adjusts with each solution and should match exactly your graphical solution.

e. Be Awesome!

    Make your own single-stage solvent extraction simulator using the template in Excel "Solvent Extraction/1-Stage DIY." Make sure you get the same answers as above.

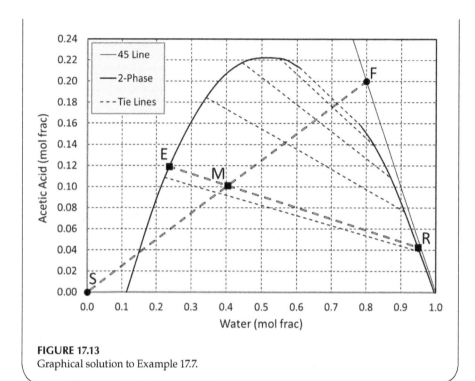

FIGURE 17.13
Graphical solution to Example 17.7.

TRY THIS AT HOME 17.3   LLE for Water–Acetic Acid–MTBE

Miao, et al. (2007) measured the Water–Acetic Acid–MTBE (methyl *tert*-butyl ether) system LLE at 25°C. MTBE has several advantages over other extraction solvents: more selective, lower density, lower vaporization enthalpy, and relatively cheap. The tie lines expressed in mass fraction are shown in Table 17.3 and Excel "MTBE Extraction/Data DIY."

1. Using Excel "MTBE Extraction/Data DIY," convert the mass fractions to mole fractions. Compare to Excel "MTBE Extraction/Data."
2. Make a professional quality graph of the LLE data showing the extract and raffinate curves and the tie lines.
3. Add a fourth-order polynomial trend line to the extract and raffinate.
4. Calculate the slope of each tie line and plot the slope as a function of the extract water mole fraction. Add a

TABLE 17.3

Mass Fractions of Water (W)–Acetic Acid (A)–MTBE (M) LLE at 25°C

Extract			Raffinate		
W	A	M	W	A	M
0.0116	0.0000	0.9884	0.9548	0.0000	0.0452
0.0259	0.0463	0.9278	0.9062	0.0476	0.0462
0.0393	0.1123	0.8484	0.8384	0.1071	0.0545
0.0721	0.1793	0.7486	0.7899	0.1524	0.0577
0.1068	0.2554	0.6378	0.6950	0.2251	0.0799
0.1620	0.2937	0.5443	0.6253	0.2686	0.1061
0.2357	0.3300	0.4343	0.5302	0.3174	0.1524
0.3065	0.3343	0.3592	0.4505	0.3248	0.2247
0.3341	0.3372	0.3287	0.4247	0.3281	0.2472

fourth-order polynomial trend line to this curve. Compare your finished ternary phase diagram to Excel "MTBE Extraction/Ternary Diagram."

The ternary phase diagram is shown in Figure 17.14. Note that the tie line compositions do not fit any curve perfectly – especially the extract side – as can be expected for real experimental data.

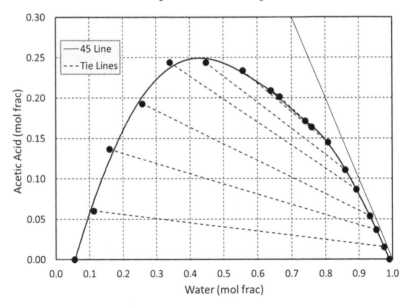

FIGURE 17.14
Ternary phase diagram for the water/acetic acid/MTBE system at 25°C.

TRY THIS AT HOME 17.4   Single-Stage Extraction Using MTBE

A total of 100 mol/s of a liquid mixture containing 90 mol% water and 10 mol% acetic acid is contacted with 75 mol/s of pure MTBE in a single-stage extractor operating at 25°C. Determine the flow rate and composition of the extract and raffinate streams. Use Excel "MTBE Extraction/DIY" to construct the simulation.

   a. Draw a labeled process flow diagram.

      The labeled PFD is shown in Figure 17.15. Fill in all known values and leave the rest blank for now.

   b. Conduct a degree of freedom analysis.

     Unknowns (6):

     Equations (6):

     There are six unknowns and six equations for zero DOF. The problem is properly specified and can be solved.

   c. Write the equations and solve.

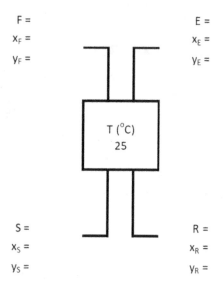

FIGURE 17.15

Unlabeled PFD for use in Try This at Home 17.4. Fill in what you know from the problem specifications, leaving the unknowns blank for now.

Overall mole balance:

Water mole balance:

Acetic acid mole balance:

Extract equation:

Raffinate equation:

Slope equation:

Input these equations into Excel "MTBE Extraction/1-Stage DIY." Use the phase diagram to make reasonable initial guesses for the compositions. You should find 93.9 mol/s of extract with 12.5 mol% water and 8.6 mol% acetic acid. Also, there will be 81.1 mol/s of raffinate with 96.5 mol% water and 2.3 mol% acetic acid. You can use Excel "MTBE Extraction/1-Stage" to check your answer. The graphical solution is shown in Figure 17.16.

    d. Explore the problem.

        Try a range of solvent flow rates. Compare the purity of the raffinate water to the amount of water lost in the extract

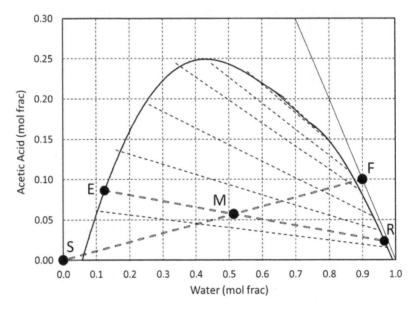

FIGURE 17.16
Graphical solution for Try This at Home 17.4

and the MTBE lost in the raffinate. Most solvent extraction processes require further processing to recover the solvent and purify the products.

e. Be awesome!

In the next section, we use the NRTL equation to model the phase equilibrium. Even if your professor considers this an optional topic, no one will stop you from trying this.

17.4 NRTL Equation

At equilibrium, the mole fractions of each component in the extract and raffinate phases are related by

$$\gamma_i^E x_i^E = \gamma_i^R x_i^R \tag{17.7}$$

where γ is the activity coefficient and is different for each component and phase. These liquids are highly non-ideal; that is why they separate into two phases! So if we can't assume $\gamma = 1$, how is it calculated? For LLE it is usually calculated by the NRTL or UNIQUAC models by fitting their parameters to experimental data. Let's just look at the "simpler" NRTL equation.

We'll start with the equation solved for the activity coefficient (not a trivial derivation from the original form for the excess Gibbs free energy!). It is usually written as a general equation for n components with nested summations. Fortunately, Walas (1985) writes it out in expanded form for three components as

$$
\begin{aligned}
\ln \gamma_i = {}& \frac{x_1 \tau_{1i} G_{1i} + x_2 \tau_{2i} G_{2i} + x_3 \tau_{3i} G_{3i}}{x_1 G_{1i} + x_2 G_{2i} + x_3 G_{3i}} \\
& + \frac{x_1 G_{i1}}{x_1 G_{11} + x_2 G_{21} + x_3 G_{31}} \left[\tau_{i1} - \frac{x_1 \tau_{11} G_{11} + x_2 \tau_{21} G_{21} + x_3 \tau_{31} G_{31}}{x_1 G_{11} + x_2 G_{21} + x_3 G_{31}} \right] \\
& + \frac{x_2 G_{i2}}{x_1 G_{12} + x_2 G_{22} + x_3 G_{32}} \left[\tau_{i2} - \frac{x_1 \tau_{12} G_{12} + x_2 \tau_{22} G_{22} + x_3 \tau_{32} G_{32}}{x_1 G_{12} + x_2 G_{22} + x_3 G_{32}} \right] \\
& + \frac{x_3 G_{i3}}{x_1 G_{13} + x_2 G_{23} + x_3 G_{33}} \left[\tau_{i3} - \frac{x_1 \tau_{13} G_{13} + x_2 \tau_{23} G_{23} + x_3 \tau_{33} G_{33}}{x_1 G_{13} + x_2 G_{23} + x_3 G_{33}} \right]
\end{aligned}
\tag{17.8}
$$

where

$$G_{ij} = \exp\left(-\alpha_{ij} \tau_{ij}\right) \tag{17.9}$$

The parameter α_{ij} is a constant related to the non-randomness of the mixture. In the collection of LLE data by Sorensen and Arlt (1980) it is always assigned a value of 0.2. The parameter $\tau_{i,j}$ is a function of the interaction energies between two molecules. Sorensen and Arlt define it as

$$\tau_{ij} \equiv A_{ij}/T(K) \qquad (17.10)$$

The binary interaction parameters (BIPs) are given as A_{ij} and A_{ji} for each combination of chemicals for a total of six BIPs. These are obtained by fitting the equation to experimental data. The LLE Data Collection is in volume V, parts 1–3, of the Chemistry Data Series published by DECHEMA and often simply referred to by the publisher.

Before we go further, a word of caution. Sources of data define terms in different and sometimes confusing ways. For instance, in DECHEMA the parameters are listed as A_{ij} but the equation is in terms of a_{ij} and it is not obvious to a novice that they should be divided by the absolute temperature. The process simulator ChemCAD uses the symbol B_{ij} for the BIPs, even though the data comes from DECHEMA. It is also extremely easy to make a mistake because of the many subscripted parameters. In fact, the expanded form in Walas has a couple mistakes that I (hopefully!) corrected in Equation 17.8. The bottom line is that you need to be very careful when using the NRTL equation.

The NRTL equation is not fundamental and is best described as a semi-empirical fit of the data. It is difficult to use as seen above and very sensitive to the initial guess for composition. Also, the totally empirical polynomial equations seem to fit the data better. In my opinion, for our purposes the polynomial equations are a better way to use the experimental data than the more conventional NRTL equation. The only limitation is not being able to extend the data to other temperatures. But let's give the NRTL equation a try!

EXAMPLE 17.8   Using the NRTL Equations to Generate LLE Graphs

Excel "NRTL/LLE" has the NRTL parameters from DECHEMA for the water–acetic acid–isopropyl acetate system at 25°C. The activity coefficient for each component to use in Eq. 17.7 is calculated using the NRTL equation (Eq. 17.8). The composition of water in the extract is specified and the composition of acetic acid in the extract and both water and acetic acid in the raffinate are adjusted until Eq. 17.7 is satisfied for all three components using Solver.

The solution is very sensitive to your initial guess for the compositions. Use the experimental ternary phase diagram to make those

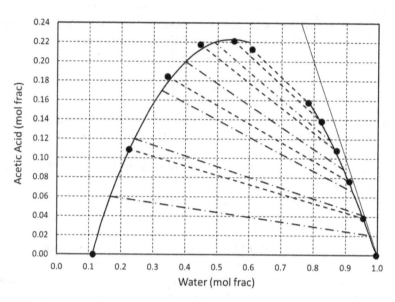

FIGURE 17.17
Water–acetic acid–isopropyl acetate liquid–liquid ternary phase diagram generated by the NRTL equation (solid line) compared to experimental data (markers). Tie lines are also shown (dash-dot for the NRTL points and dash for the experimental data).

initial guesses. This was done for $x_E = 0.112$–0.55 and the resulting other compositions manually transferred to the table headed by "LLE Data" in Excel "NRTL/LLE."

The experimentally measured compositions are plotted in Figure 17.17. The NRTL equation (solid line) has fit the experimental data reasonably well, especially for the raffinate, but not as well as the simple fourth-order polynomial equation in Figure 17.3. Tie lines are also plotted for the experimental compositions (dashed lines) and the NRTL equation (dash-dot lines).

EXAMPLE 17.9   Single-Stage Solvent Extraction Using NRTL Equations

A total of 100 mol/s of a liquid feed containing 75 mol% water and 25 mol% acetic acid is to be extracted with 100 mol/s of solvent containing 95% isopropyl acetate and 5 mol% acetic acid. The process is carried out at 25°C and 1,000 mm Hg. Determine the flow rates and compositions of the raffinate and extract products.

a. Draw a labeled process flow diagram and

b. Conduct a degree of freedom analysis.

A labeled PFD is shown in Figure 17.18. We have seen before that completely specifying the feed and solvent flow rates and compositions yields zero DOF.

c. Write the equations and solve.

Overall mole balance: $F + S = E + R$

Water mole balance: $Fx_F = Ex_E + Rx_R$

Acetic acid mole balance: $Fy_F + Sy_S = Ey_E + Ry_R$

Water phase equilibrium equation: $\gamma_W^E x_E = \gamma_W^R x_R$

Acetic acid phase equilibrium equation: $\gamma_A^E y_E = \gamma_A^R y_R$

Isopropyl acetate phase equilibrium equation: $\gamma_I^E (1 - x_E - y_E) = \gamma_I^R (1 - x_R - y_R)$

The activity coefficients from the NRTL equations and the solution to this example are calculated in Excel "NRTL/1-Stage." The solution is shown in Table 17.4 along with the solution with the polynomial equations. A graphical representation of the solution is shown in Figure 17.19. Fill in remaining flow rates and compositions in Figure 17.18.

d. Explore the problem.

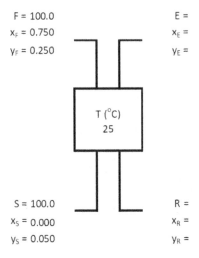

F = 100.0   E =
x_F = 0.750 x_E =
y_F = 0.250 y_E =

T (°C)
25

S = 100.0   R =
x_S = 0.000 x_R =
y_S = 0.050 y_R =

FIGURE 17.18
Process flow diagram for the solvent extraction process described in Example 17.9.

TABLE 17.4

Comparison of Results: NRTL versus Polynomial Fit

NRTL	Polynomial
$E = 178$	$E = 176$
$y_W = 0.31$	$y_W = 0.30$
$y_A = 0.16$	$y_A = 0.16$
$R = 22$	$R = 24$
$x_W = 0.93$	$x_W = 0.93$
$x_A = 0.06$	$x_A = 0.06$

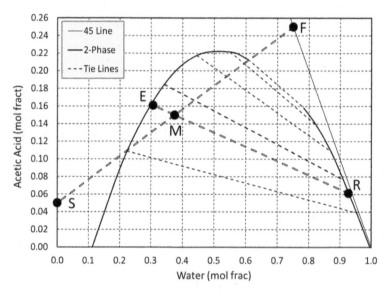

FIGURE 17.19

Graphical representation of the solution to Example 17.9 using the NRTL equations for phase equilibrium. The LLE curve and tie lines are from experimental data.

Try different combinations of solvent and feed flow rates and compositions. Compare the solutions to Excel "Solvent Extraction/1-Stage" that uses the polynomial equations and the resulting graphical solution in Excel "Solvent Extraction/1-Stage Graph." If the solution doesn't look correct, try a better starting guess for the compositions.

e. Be awesome!

Try inputting the NRTL equations into the spreadsheet Excel "NRTL/DIY" yourself. If you succeed, you might be awesome – or you might just have too much time on your hands!

17.5 Summary

In the solvent extraction process, a solute dissolved in a diluent is removed by contacting with solvent. The diluent and solvent must be at least partially immiscible to allow physical separation of the products. Raffinate is the liquid product with most of the diluent, while extract is the liquid product with most of the solvent and solute. The composition and amount of each product is determined by mole balances, phase equilibrium relationships, and various process specifications. These equations can be solved graphically or analytically. Measured compositions are best represented by a ternary phase diagram. The phase equilibrium equations can be purely empirical by fitting measured compositions to polynomial equations, or semi-empirical by fitting measured compositions to a model such as NRTL or UNIQUAC.

18

Mathematical Analysis of Solvent Extraction Columns

We found that separation processes – distillation, absorption, and stripping – always produced better separation using multiple stages compared to a single stage. It is the same with solvent extraction. The process will look very much like stripping in which a liquid feed is contacted with a gas, except the gas is replaced by a second, immiscible liquid. The leaving and passing streams are related by the same equations: mole balances, phase equilibrium relationships, and various process specifications. There are simply more – but not different – equations.

18.1 Two Parallel Stages

We can examine two stages in two configurations shown in Figure 18.1. In the first, the solvent is split in half and used to extract the feed twice, discarding the extract each time. We'll call that a parallel process from the perspective of the solvent. In the second, the feed is contacted in Stage 1 by the

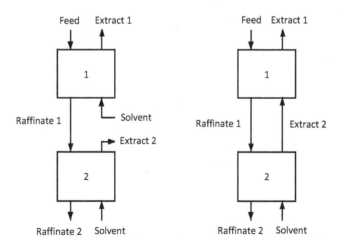

FIGURE 18.1
Configurations for two-stage solvent extraction process.

extract from Stage 2 and then the raffinate from Stage 1 is contacted with the full amount of fresh solvent in Stage 2. The latter is equivalent to the use of reflux in distillation. We'll call this a countercurrent process. It will be best to demonstrate this with a concrete example.

EXAMPLE 18.1   Solvent Extraction in Two Parallel Stages

A total of 100.0 mol/s of 20 mol% acetic acid in water is contacted with 15.0 mol/s of pure isopropyl acetate in a mixer-settler. The two liquids are mixed by vigorous agitation, allowed to settle in an unmixed tank, and drained to separate tanks. The process is continued with an additional 15.0 mol/s of isopropyl acetate added to the water-rich raffinate in a second mixer-settler. Determine the amount and composition of each phase after each step. Note that extracting 100.0 mol/s of a 20 mol% acetic acid in water feed with a total of 30.0 mol/s of pure solvent will be used in all configurations of multiple extraction stages in this chapter so we can compare results. We can also compare these results to a single-stage extraction with 30 mol/s of solvent.

a. Draw a labeled process flow diagram.
    The labeled process flow diagram (PFD) is shown in Figure 18.2.

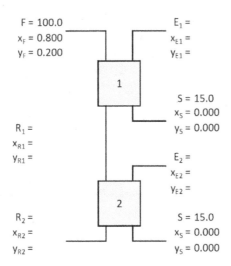

FIGURE 18.2
Process flow diagram of a two-stage extraction process in parallel. Flow rates (E, F, S, R) are in units of mol/s and compositions (x, y) are in units of mole fractions. Process conditions were specified in Example 18.1.

b. Conduct a degree of freedom analysis.

Stage 1

Unknowns (6): E_1, x_{E1}, y_{E1}, R_1, x_{R1}, y_{R1}

Equations (6):

3 mole balances (total, water, acetic acid)

3 phase equilibrium relationships (extract, raffinate, and tie line slope equations)

We will use the polynomial functions for the extract, raffinate, and slope curves to describe the equilibrium compositions. There are zero degrees of freedom (DOF) and the problem can be solved. The second extraction step would have three DOF so the problem must be solved in order. (Alternatively, both stages can be solved simultaneously.)

c. Write the equations and solve.

Overall mole balance: $F + S = R_1 + E_1$

Water mole balance: $Fx_F + Sx_S = R_1x_{R1} + E_1x_{E1}$

Acetic acid mole balance: $Fy_F + Sy_S = R_1y_{R1} + E_1y_{E1}$

Extract equation: $y_{E1} = a_E x_{E1}{}^4 + b_E x_{E1}{}^3 + c_E x_{E1}{}^2 + d_E x_{E1} + e_E$

Raffinate equation: $y_{R1} = a_R x_{R1}{}^4 + b_R x_{R1}{}^3 + c_R x_{R1}{}^2 + d_R x_{R1} + e_R$

Slope equation: $(y_{E1} - y_{R1})/(x_{E1} - x_{R1}) = a_S x_{E1}{}^4 + b_S x_{E1}{}^3 + c_S x_{E1}{}^2 + d_S x_{E1} + e_S$

These equations can be solved using Excel "Solvent Extraction/2-Stage Parallel." You should find 52.3 mol/s of extract with composition (0.530, 0.223) and 62.7 mol/s of raffinate with composition (0.834, 0.133) after the first extraction. Note that the raffinate is now contaminated with 2.1 mol/s of isopropyl acetate and that we lost 27.7 mol/s of water (out of 80.0 mol/s of water) to the extract.

The spreadsheet is set to solve the two extraction steps simultaneously (the "brute force" method). You can easily modify the Solver settings to solve the first stage and then the second stage. Either way, 46.1 mol/s of raffinate with composition (0.923, 0.066) is achieved after the second step. There is an increase in water purity from 89.3 mol% in the single-stage extraction with 30.0 mol/s of solvent using Excel "Solvent Extraction/1-Stage." However, we lost 37.4 mol/s of water to the two extract streams compared to 28.1 mol/s in the single stage. Fill in the remaining flow rates and compositions in Figure 18.2. Check by hand calculations that these flow rates and compositions satisfy the equations.

d. Explore the problem.

Using Excel "Solvent Extraction/2-Stage Parallel," systematically increase the amount of solvent used in each step. For each amount of solvent, make a table, showing the raffinate and extract compositions, and amount of water lost to the extract. Presumably, the purpose of this procedure is to purify the water – the more solvent used, the purer the water. This comes at a cost, though, of losing more water to the extract.

e. Be awesome!

Add a third step to Excel "Solvent Extraction/2-Stage Parallel" and compare the performance to one- and two-stage processes. Keep the total amount of solvent constant by dividing the one-stage process solvent by two for the two-stage and by three for the three-stage. Compare performances.

TRY THIS AT HOME 18.1   Two Parallel Stages Solved Graphically

The problem in Example 18.1 can be solved graphically. Print the ternary phase diagram from Excel "Solvent Extraction/Ternary Diagram." Follow my lead:

1. Locate the feed (0.80, 0.20) and the solvent (0.00, 0.00) compositions and draw a line between them. Locate the mix point using a simple mole balance or the inverse lever rule (even better – using both methods!). The mix point will always be on the connecting line and in this case is approximately (0.69, 0.17).
2. Draw a line between the extract and raffinate curves that intersects the mix point by interpolating between tie lines.
3. Locate the compositions of the raffinate and extract – approximately (0.83, 0.13) and (0.53, 0.22), respectively.
4. Calculate the amount of raffinate and extract using the inverse lever rule or a simple mole balance (even better – using both methods!) – approximately 63 and 52 mol/s, respectively.
5. The point on the raffinate curve now becomes the feed to the second extraction step. Repeat steps 1–4 with a fresh aliquot of isopropyl acetate. Your results should be reasonably close to the analytical solution. Compare your graph to Figure 18.3.

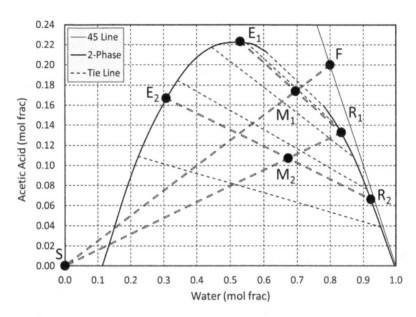

FIGURE 18.3
Ternary phase diagram with compositions of the feed (F), solvent (S), extract (E), raffinate (R), and mixtures (M) identified for two solvent extractions. Process conditions were specified in Example 18.1. The graphical solution was described in Try This at Home 18.1.

18.2 Two Countercurrent Stages

In a staged column or multiple mixer-settlers, this separation can be carried out by using the extract from the second stage as the solvent for the first (instead of just throwing it out). In Chapter 5, we called this reflux or recycle. It will not be obvious yet how to conduct a graphical solution so only the mathematical solution will be attempted for now.

EXAMPLE 18.2 Solvent Extraction in Two Countercurrent Stages

A total of 100 mol/s of 20 mol% acetic acid in water is contacted with 30 mol/s of pure isopropyl acetate in a countercurrent column with two stages. The aqueous stream is fed to the top stage and the solvent is fed to the bottom stage. Determine the amount and composition of every leaving and passing stream.

a. Draw a labeled process flow diagram.

The labeled PFD is shown in Figure 18.4. Note that this design requires a significant density difference between the raffinate and extract phases. In a column, mixing can be enhanced by pressure pulsing.

b. Conduct a degree of freedom analysis.

For Stage 1:

Unknowns (9): R_1, x_{R1}, y_{R1}, E_1, x_{E1}, y_{E1}, E_2, x_{E2}, y_{E2},

Equations (6):

   3 mole balances (total, water, acetic acid)

   3 phase equilibrium relationships (raffinate, extract, and tie line slope equations)

We will use the polynomial functions for the extract, raffinate, and slope curves to describe the equilibrium compositions. There are three DOFs so the problem cannot be solved. The second stage would also have three DOFs. But the passing streams between Stages 1 and 2 are shared so taking the whole

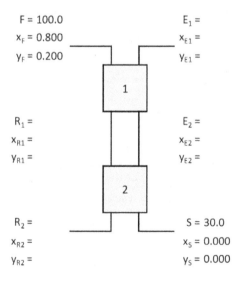

$F = 100.0$ $E_1 =$

$x_F = 0.800$ $x_{E1} =$

$y_F = 0.200$ $y_{E1} =$

1

$R_1 =$ $E_2 =$

$x_{R1} =$ $x_{E2} =$

$y_{R1} =$ $y_{E2} =$

2

$R_2 =$ $S = 30.0$

$x_{R2} =$ $x_S = 0.000$

$y_{R2} =$ $y_S = 0.000$

FIGURE 18.4
Process flow diagram of a two-stage extraction process in series. Flow rates (E, F, S, R) are in units of mol/s and compositions (x, y) are in units of mole fractions. Process conditions were specified in Example 18.2.

process, including interior streams, results in 12 unknowns and 12 equations (three mole balances and three phase equilibrium relationships for each stage) and zero DOF. Unlike the parallel stages, a simultaneous solution must be made for countercurrent stages.

c. Write the equations and solve.

Stage 1

Overall mole balance: $F + E_2 = R_1 + E_1$

Water mole balance: $Fx_F + E_2x_{E2} = R_1x_{R1} + E_1x_{E1}$

Acetic acid mole balance: $Fy_F + E_2y_{E2} = R_1y_{R1} + E_1y_{E1}$

Extract equation: $y_{E1} = a_Ex_{E1}^4 + b_Ex_{E1}^3 + c_Ex_{E1}^2 + d_Ex_{E1} + e_E$

Raffinate equation: $y_{R1} = a_Rx_{R1}^4 + b_Rx_{R1}^3 + c_Rx_{R1}^2 + d_Rx_{R1} + e_R$

Slope equation: $(y_{E1} - y_{R1})/(x_{E1} - x_{R1}) = a_Sx_{E1}^4 + b_Sx_{E1}^3 + c_Sx_{E1}^2 + d_Sx_{E1} + e_S$

Stage 2

Overall mole balance: $R_1 + S = R_2 + E_2$

Water mole balance: $R_1x_{R1} + Sx_S = R_2x_{R2} + E_2x_{E2}$

Acetic acid mole balance: $R_1y_{R1} + Sy_S = R_2y_{R2} + E_2y_{E2}$

Extract equation: $y_{E2} = a_Ex_{E2}^4 + b_Ex_{E2}^3 + c_Ex_{E2}^2 + d_Ex_{E2} + e_E$

Raffinate equation: $y_{R2} = a_Rx_{R2}^4 + b_Rx_{R2}^3 + c_Rx_{R2}^2 + d_Rx_{R2} + e_R$

Slope equation: $(y_{E2} - y_{R2})/(x_{E2} - x_{R2}) = a_Sx_{E2}^4 + b_Sx_{E2}^3 + c_Sx_{E2}^2 + d_Sx_{E2} + e_S$

The equations are solved in Excel "Solvent Extraction/2-Stage." Figure 18.5 shows the compositions of all streams. The two leaving streams (related by phase equilibrium) and three passing streams (related by mole balances) are connected by lines. Notice that performance is significantly improved over a single stage and two parallel stages, even when using the same amount of solvent. Fill in the calculated flow rates and compositions into Figure 18.4 and verify by hand calculations that the equations are satisfied.

d. Explore the problem.

Using Excel "Solvent Extraction/2-Stage," systematically increase the amount of solvent used. Complete the table below showing the product compositions and flow rates. Discuss the results with your study partners.

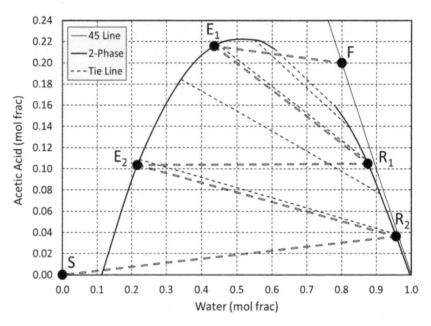

FIGURE 18.5

Ternary phase diagram with compositions of the feed (F), solvent (S), extract (E), raffinate (R), and mixtures (M) identified for two series solvent extractions. Process conditions were specified in Example 18.2.

		Raffinate			Extract	
Solvent (mol/s)	Moles	Water	Acetic Acid	Moles	Water	Acetic Acid
15						
20						
25						
30						

e. Be awesome!

Make your own simulation for two countercurrent stages by completing the equations in Excel "Solvent Extraction/2-Stage DIY." Instead of specifying the solvent flow rate, leave S as an unknown. Add the process specification that a percentage of the acetic acid entering with the feed is removed with the extract. Try this with 90%, 95%, and 99% acetic acid removed and plot solvent flow rate versus purity. If you try this with 10 mol/s solvent or 85% acetic acid removal, the simulation doesn't work. Why? Discuss with a classmate. (Hint: Are you still in the two-phase region?)

18.3 Three Countercurrent Stages

Now we will look at three countercurrent stages and continue to use the "brute force" method to solve for unknown flow rates and compositions. Already the solution begins to tax the capabilities of Excel's Solver tool making a stage-by-stage equivalent of the Lewis method difficult.

TRY THIS AT HOME 18.2   Solvent Extraction in Three Stages

A total of 100 mol/s of 20 mol% acetic acid in water is contacted with 30 mol/s of isopropyl acetate in a column with three stages. Determine the amount and composition of every leaving and passing stream.

a. Draw a labeled process flow diagram.
   The PFD is shown in Figure 18.6. Fill in any specified values and leave the rest blank for now.

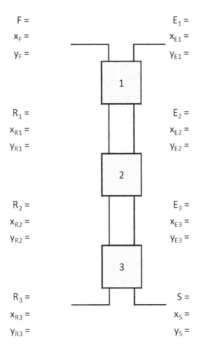

$F =$		$E_1 =$
$x_F =$		$x_{E1} =$
$y_F =$		$y_{E1} =$

$R_1 =$		$E_2 =$
$x_{R1} =$		$x_{E2} =$
$y_{R1} =$		$y_{E2} =$

$R_2 =$		$E_3 =$
$x_{R2} =$		$x_{E3} =$
$y_{R2} =$		$y_{E3} =$

$R_3 =$		$S =$
$x_{R3} =$		$x_S =$
$y_{R3} =$		$y_S =$

FIGURE 18.6
Process flow diagram of a three-stage extraction process in series. Flow rates (E, F, S, R) are in units of mol/s and compositions (x, y) are in units of mole fractions. Process conditions were specified in Try This at Home 18.2. Fill in specified values and first, solve the problem, then fill in the rest.

b. Conduct a degree of freedom analysis.

   The DOF analysis will be similar to the one in Example 18.2. There is no part of this system with zero DOF. However, there will be three mole balances and three phase equilibrium equations (the polynomials again) for each stage, amounting to 18 independent equations. There are also, 18 unknown flow rates and compositions (be sure to verify using the PFD in Figure 18.6). So solving everything together will have zero DOF.

c. Write the equations and solve.

   Stage 1
   Overall mole balance:

   Water mole balance:

   Acetic acid mole balance:

   Extract equation:

   Raffinate equation:

   Slope equation:

   Stage 2
   Overall mole balance:

   Water mole balance:

   Acetic acid mole balance:

   Extract equation:

   Raffinate equation:

   Slope equation:

   Stage 3
   Overall mole balance:

   Water mole balance:

   Acetic acid mole balance:

Extract equation:

Raffinate equation:

Slope equation:

Input these equations into Excel "Solvent Extraction/3-Stage DIY." Make reasonable guesses for the unknown flow rates and compositions and then run Solver. Fill in all values into Figure 18.6. If you have trouble with this, the problem is solved in Excel "Solvent Extraction/3-Stage." Figure 18.7 shows the stream compositions and lines connecting passing and leaving streams for each stage.

Solver has trouble converging on a solution for this many equations if the guessed values are too far off. Try these "intelligent" guesses. The raffinate will lose flow rate as it progresses through the column, so guess $R_1 = 80$, $R_2 = 60$, and $R_3 = 40$. The compositions will be much purer in water, so guess $x_{R1} = 0.85$, $x_{R2} = 0.90$, $x_{R3} = 0.95$, $y_{R1} = 0.15$, $y_{R2} = 0.10$, and $y_{R3} = 0.05$. The extract will gain flow rate as it progresses through the column, so guess $E_1 = 50$, $E_2 = 70$, and $E_3 = 90$. The extract becomes more concentrated in water and acetic acid, so guess $x_{E1} = 0.10$, $x_{E2} = 0.20$, $x_{E3} = 0.30$, $y_{E1} = 0.10$, $y_{E2} = 0.20$, and $y_{E3} = 0.30$. Now run Solver for the exact solution.

d. Explore the problem.

The table below shows the raffinate and extract amounts and compositions for each of the four systems studied. The one-stage process was calculated in Example 18.1. Fill in the values and compare performances.

Process	Raffinate			Extract		
	Moles	Water	Acetic Acid	Moles	Water	Acetic Acid
1-Stage						
2-Stage parallel						
2-Stage						
3-Stage						

Suppose you are performing a preliminary design analysis of this process for your boss, Dr. P. H. Dee at BamaCHEM. Write a short report on your results and discuss the tradeoffs in capital, material, and utility costs between the various options.

e. Be awesome!

A graphical solution for multiple solvent extraction stages in series was not immediately obvious because the interior stream compositions were not known. Print out Figure 18.7 (from Excel "Solvent Extraction/3-Stage Graph") and tape several sheets of paper to the right side of the graph as seen in Figure 18.8.

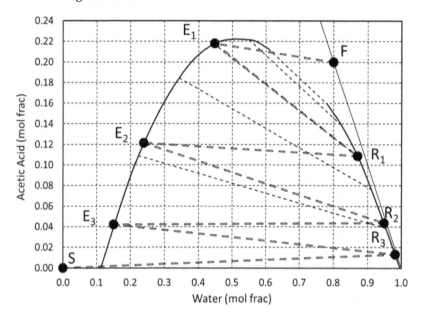

FIGURE 18.7
Ternary phase diagram with compositions of the feed (F), solvent (S), extract (E), raffinate (R), and mixtures (M) identified for three series solvent extractions. Process conditions were specified in Try This at Home 18.2.

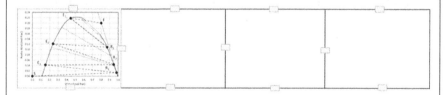

FIGURE 18.8
An example of taping blank sheets to a ternary phase diagram for use in graphical analysis. This is too small to use effectively. Instead, print Figure 18.7 from Excel "Solvent Extraction/3-Stage Graph" and tape three blank sheets to it.

Notice that the passing stream lines look like they converge somewhere on the right. Take a long straight edge and extend those lines to the right. If you are very careful with a steady hand and good eyesight, you should find that they all converge on the same spot! Curious!

Congratulations! You have just discovered a graphical technique for multiple solvent extraction stages called the Hunter-Nash method. The method and its algebraic basis will be discussed in the next chapter.

Back in the day (I entered college in 1971 with a slide rule strapped to my belt!) chemical engineering professors could be much more cruel than today. We would get situations where those passing stream lines were nearly parallel and wouldn't intersect for quite a distance. We would start at one end of the hall on the floor with a large hand-drawn graph (personal computers would not be around for many years) and tape down a roll of butcher paper. Then we would pin one end of string to the extract point and extend the line down the hall. This would be done for each set of passing streams until the lines intersected. You can do this in minutes on a computer by simply increasing the range on the x-axis. Check out Excel "Solvent Extraction/3-Stage Hunter-Nash."

18.4 Summary

Using multiple equilibrium stages increases the amount of solute transferred to the solvent. The effect is greater with countercurrent stages than in parallel. The analysis requires the same equations (mole balances, phase equilibrium equations, and various process specifications) as a single stage, just more of them. As in distillation, absorption, and stripping, if the passing and leaving streams are plotted on a phase diagram, a graphical technique becomes evident.

19

Graphical Design of Solvent Extraction Columns

In Chapter 6, we saw that mathematical analysis could determine the performance of a distillation column given the design. Graphical methods (McCabe–Thiele) in Chapter 7 could determine the design given the desired performance. The same was true for absorption/stripping and will be in this chapter for solvent extraction.

19.1 Graphical Method

In Chapter 18, we observed that the lines connecting passing stream compositions (x_{Ei}, y_{Ei}), $(x_{R(i-1)}, y_{R(i-1)})$ all converged on the same spot when extended beyond the ternary phase diagram. If we can determine the compositions of two sets of passing streams, that intersection can be found. The top and bottom streams can be used after performing an overall mole balance.

Figure 19.1 is a solvent extraction column process flow diagram (PFD) consisting of an unknown number of stages. Usually the feed and solvent streams are specified and these compositions can be located. If we know their flow rates and the number of stages that is sufficient information to calculate all other stream flow rates and compositions (an analysis problem). If we don't know the number of stages (a design problem) then performance must be specified, such as mole fraction of solute in the raffinate or mol% of solute extracted.

At first look, there still seems to be not enough information to determine the top and bottom passing streams from an overall mole balance. The final raffinate and extract are not in equilibrium, but their compositions do lie on the two-phase line, so if you know one mole fraction you know the others. If x_{RN} is specified, for example, the five unknowns are R_N, y_{RN}, E_1, x_{E1}, and y_{E1}. The five equations to solve for these include three mole balances and two equilibrium relationships – (x_{RN}, y_{RN}) is on the raffinate curve and (x_{E1}, y_{E1}) is on the extract curve. Of course, in this graphical method, E_1 will be found from the mixing rule.

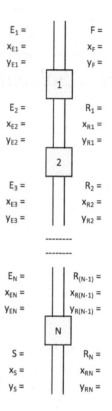

FIGURE 19.1
An N-stage solvent extraction column. The molar flow rates of each stream (F, S, EN, and RN) are shown with their compositions as mole fractions with x being water and y being acetic acid.

While leaving streams are related by phase equilibrium tie lines, passing streams are related by mole balances. Draw a material balance envelope (MBE) around Stage 1 in Figure 19.1. The total mole balance is

$$F + E_2 = R_1 + E_1 \qquad (19.1)$$

Now draw an MBE envelope around Stages 1 and 2. The total mole balance is

$$F + E_3 = R_2 + E_1 \qquad (19.2)$$

Collecting passing streams on one side of the equation, it is easy to see that

$$F - E_1 = R_1 - E_2 = R_2 - E_3 = R_i - E_{i+1} = R_N - S \equiv D \qquad (19.3)$$

D is defined to be the difference between passing streams and is called the difference point.

Mole balances for each component can also be written

$$Fx_F + E_2 x_{E2} = R_1 x_{R1} + E_1 x_{E1} \tag{19.4}$$

$$Fy_F + E_2 y_{E2} = R_1 y_{R1} + E_1 y_{E1} \tag{19.5}$$

and expressed as differences in passing streams.

$$Fx_F - E_1 x_{E1} = R_1 x_{R1} - E_2 x_{E2} = R_i x_{Ri} - E_{i+1} x_{E(i+1)}$$
$$= R_N x_{RN} - S x_S \equiv D x_D \tag{19.6}$$

$$Fy_F - E_1 y_{E1} = R_1 y_{R1} - E_2 y_{E2} = R_i y_{Ri} - E_{i+1} y_{E(i+1)}$$
$$= R_N y_{RN} - S y_S \equiv D y_D \tag{19.7}$$

The coordinates of the difference point are (x_D, y_D). Solving for these yields

$$x_D = \frac{Fx_F - E_1 x_{E1}}{D} = \frac{R_N x_{RN} - S x_S}{D} \tag{19.8}$$

$$y_D = \frac{Fy_F - E_1 y_{E1}}{D} = \frac{R_N y_{RN} - S y_S}{D} \tag{19.9}$$

So you can use either the top or bottom passing streams to calculate the difference point coordinates. You can also find the difference point graphically by extending the line between F and E_1 (the top operating line (TOL)) beyond the graph (need to tape on extra paper!) and doing the same for the line between S and R_N (the bottom operating line (BOL)). The intersection of these lines is the difference point. Notice that Eqs. 19.6 and 19.7 can be written as a mixing calculation (as was done for the overall process) between points E_1 and D. Using the x-axis as a ruler, for instance

$$Dx_D + E_1 x_{E1} = Fx_F \tag{19.10}$$

or more generally

$$Dx_D + E_{i+1} x_{E(i+1)} = R_i x_{Ri} \tag{19.11}$$

Since these points will be colinear, i.e., x_D, $x_{E(i+1)}$, and x_{Ri} are on a straight line, if you know x_D and the extract composition, you can find the raffinate composition and vice versa.

This graphical technique is called the Hunter-Nash method (Hunter and Nash, 1934). The general procedure for graphical design of a solvent extraction column using a ternary phase diagram is:

1. Determine the flow rate and composition of the feed (F), solvent (S), final raffinate (R_N), and final extract (E_1) using the graphical mixing rule and locate them on the diagram.

2. Extend a line through the passing stream compositions (F and E_1, S and R_N) until they intersect outside the diagram. These are the TOL and BOL. Sometimes they intersect on the left, sometimes on the right, and if you are the unlucky type, are nearly parallel. That intersection is the difference point.

3. Draw an equilibrium (tie) line from the extract, point E_1, to the raffinate, R_1. This will likely require interpolation between experimental tie lines.

4. Draw an operating line from point D to point R_1 and then extend the straight line to the extract curve. The intersection is point E_2.

5. Repeat steps 3 and 4 until the process specification is met. As with the McCabe–Thiele method, it would be coincidental if you "land" exactly on the process specification. More likely, you will want to overshoot it and err with a conservative estimate.

As always, this is best demonstrated with a worked example.

EXAMPLE 19.1   Graphical Design of a Solvent Extraction Column

A total of 100 mol/s of a mixture with 20 mol% acetic acid and 80 mol% water is extracted with 30 mol/s of pure isopropyl acetate in a multi-stage column operating at 1 atm and 30°C. The aqueous product must be at least 98 mol% water. How many equilibrium stages are required? I strongly suggest that you print out a ternary phase diagram for this system (Excel "Solvent Extraction/LLE") and follow along with the solution.

a. Draw a process flow diagram.
   The labeled PFD is drawn in Figure 19.2. At this point, we don't know how many stages will be required.

b. Conduct a degree of freedom analysis.

Overall Process

Unknowns (5): E_1, x_{E1}, y_{E1}, R_N, y_{RN}

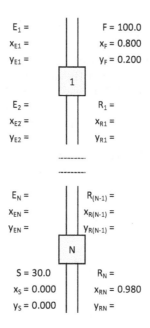

$E_1 =$ $F = 100.0$
$x_{E1} =$ $x_F = 0.800$
$y_{E1} =$ $y_F = 0.200$

1

$E_2 =$ $R_1 =$
$x_{E2} =$ $x_{R1} =$
$y_{E2} =$ $y_{R1} =$

$E_N =$ $R_{(N-1)} =$
$x_{EN} =$ $x_{R(N-1)} =$
$y_{EN} =$ $y_{R(N-1)} =$

N

$S = 30.0$ $R_N =$
$x_S = 0.000$ $x_{RN} = 0.980$
$y_S = 0.000$ $y_{RN} =$

FIGURE 19.2
Labeled PFD of the solvent extraction process described in Example 19.1.

Equations (5):

3 mole balances (one for each chemical)

2 equilibrium relationships ((x_{E1}, y_{E1}) is on the extract curve and (x_{RN}, y_{RN}) is on the raffinate curve.)

Note that streams E_1 and R_N are not in equilibrium so that they will not be on the same tie line. There are zero DOF so all flow rates and compositions of the top and bottom streams can be determined.

Stage 1
While we are looking at DOF – let's try Stage 1. Start with assuming stream E_1 has been characterized by the overall process mole balance.

Unknowns (6): E_2, x_{E2}, y_{E2}, R_1, x_{R1}, y_{R1}

Equations (6):

3 mole balances (one for each chemical)

3 equilibrium relationships ((x_{E2}, y_{E2}) is on the extract curve, (x_{R1}, y_{R1}) is on the raffinate curve, and (x_{E2}, y_{E2}) is in equilibrium (on a tie line) with (x_{R1}, y_{R1})).

There are zero DOF. This entire problem should be solvable using the Hunter–Nash method.

c. Write the equations and solve.

In Figure 19.3a, the feed (F) and solvent (S) are located at (0.80, 0.20) and (0.00, 0.00), respectively. The process raffinate (R_N) is located at $x_{RN} = 0.98$ on the raffinate curve. We need to do some calculations to know where to locate the process extract (E_1).

A line is drawn between the feed (F) and solvent (S). The mix point for the overall process is found by the inverse lever rule using the x-axis as a ruler.

$$\overline{SF} = 0.8 - 0.0 = 0.8$$

$$\frac{\overline{SM}}{\overline{SF}} = \frac{100}{130}$$

$$\overline{SM} = \frac{100}{130}0.8 = 0.62$$

The mix point (M) is located on the line at $x = 0.62$ with $y = 0.15$. These are the water and acetic acid concentrations in the mixture.

The overall process products must have the same mix point, so a line is drawn from R_N through M until it intersects with the extract curve at point E_1. The coordinates of E_1 are (0.45, 0.22). Again using the inverse lever rule and the x-axis as a ruler, the amount of E_1 and R_N can be found (although if all we want to know is the number of stages, this is not absolutely necessary).

Now we can draw the TOL and BOL in Figure 19.3b. These are constructed by drawing a line from E_1 to F and extending it to the right. (Here you should tape several sheets of paper on the right hand side (rhs) of your graph.) Draw another line from S to R_N and extend it to the right. They intersect at point D with coordinates (3.65, 0.06) if you have also extended the scale, have a steady hand, and good eyesight! It is quite difficult to do this accurately.

You can also do this by expanding the horizontal axis scale as seen in Figure 19.3c but that is a little harder to draw. In Figure 19.4, I took the phase diagram (with the top and bottom streams marked) and a blank diagram with an identical scale (Excel "Solvent Extraction/LLE and /blank"),

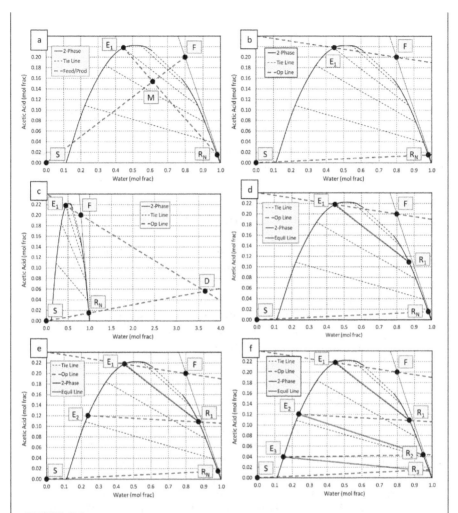

FIGURE 19.3

Demonstration of the Hunter-Nash method for graphical design of solvent extraction columns. (a) Find the column extract (E_1) using the mixing rule. (b) Draw the top and BOLs. (c) Find the difference point by extending the operating lines. (d) Find the top stage raffinate (R_1) by following the equilibrium (tie) line. (e) Find the second stage extract by drawing an operating line through D and R_1. (f) Continue alternating equilibrium (tie) line and operating line constructions until the raffinate meets or exceeds specifications. Note that after three stages, R_3 slightly overshoots the specification.

copying them onto a blank sheet of 8.5 × 14 (in MS Word), reducing them to 2.5 in wide, cropping the edges, lining them up, taping them together, and adding the operating lines. Whew! The point is, there are many ways of doing this, and I'm sure you are much cleverer than me!

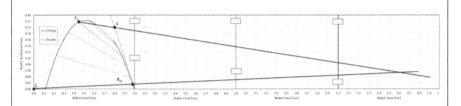

FIGURE 19.4
Demonstration of graphical method by taping blank sheets of graph paper together and extending the TOL and BOL until they intersect.

Keep in mind that graphical methods are approximate and limited by your skill at constructing the operating and equilibrium lines. In this example, I cheated by first calculating each step and then drawing the lines as if I were constructing these by hand. So don't feel bad if you can't reproduce this solution exactly.

Stream R_1 must be in equilibrium with stream E_1, so draw a tie line from E_1 to the raffinate curve. It so happens that E_1 is almost right on top of a tie line so interpolation is easy. This step is illustrated in Figure 19.3d.

Stream E_2 is a passing stream to stream R_1 and so must be on an operating line with point D. Draw a line from D to R_1 and extend it to intersect with the extract curve. This point is stream E_2. This step is illustrated in Figure 19.3e.

Continue drawing alternating equilibrium and operating lines until you achieve a raffinate that meets or exceeds specifications. In this example, R_3 has coordinates (0.98, 0.01) and in fact slightly overshoots the specification as seen in Figure 19.3f. The original BOL was drawn with R_N being precisely 0.98 mole fraction water. This system actually achieves 0.983 mole fraction water in the final raffinate as the close-up shows in Figure 19.5 – but this is a precision you cannot be sure of in a graphical solution.

The required stages and their extract and raffinate flow rates and compositions are summarized in Table 19.1.

d. Explore the problem.

The flow rates shown in Table 19.1 were determined graphically using the inverse lever rule and by mole balances. Use both methods to verify the values.

e. Be awesome!

Using more solvent should reduce the number of stages required. Print the ternary phase diagram in Excel "Solvent

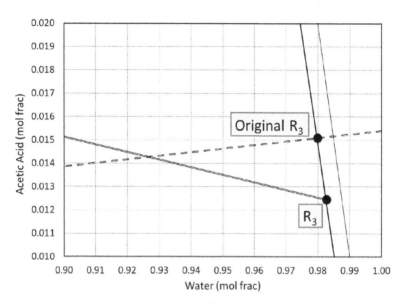

FIGURE 19.5
Close up of the graphical solution to Example 19.1.

TABLE 19.1

Flow Rates and Compositions

Stream	Flow (mol/s)	x (mol W/mol)	y (mol A/mol)
Feed	100	0.80	0.20
E_1	89	0.45	0.22
R_1	59	0.87	0.11
E_2	49	0.24	0.12
R_2	48	0.95	0.04
E_3	37	0.15	0.04
R_3	41	0.98	0.01
S	30	0.00	0.00

Extraction/LLE" and find the solvent flow rate required to achieve this specification in one stage. You can use Excel "Solvent Extraction/1-Stage" to check your graphical result. Discuss with your study partners the tradeoffs in using this much solvent.

To achieve a raffinate of 98.0 mol% water exactly in three stages would require slightly less solvent. Use Excel "Solvent Extraction/3-Stage" to determine that flow rate.

**TRY THIS AT HOME 19.1   Graphical Design
of a Solvent Extraction Process**

A total of 100 mol/s of a mixture with 25 mol% acetic acid and
75 mol% water is extracted with 30 mol/s of pure isopropyl acetate
in a multi-stage column operating at 1 atm and 30°C. The aqueous product must be at least 99 mol% water. How many equilibrium
stages are required? Print a blank phase diagram (Excel "Solvent
Extraction/LLE") and solve using the Hunter–Nash graphical
method.

a. Draw a labeled process flow diagram and

b. Conduct a degree of freedom analysis.

Other than feed composition and desired water purity, this
problem is very similar to Example 19.1. On a blank piece of
paper, sketch the labeled PFD assuming it has three stages.
As before, there are zero DOF.

c. Write the equations and solve.

We will use the Hunter-Nash graphical method. You
can check your results with the analytical solution
(Excel "Solvent Extraction/3-Stage") and graphical solution
(Excel "Solvent Extraction Graphical") afterwards.

1. Locate and mark the feed (F) and solvent (S) composi-
    tions and draw a line between them.
2. Use both a mole balance and the mixing rule to locate
    and mark the mix point.
3. Locate and mark the raffinate composition (R_N on the
    raffinate curve with $x_{RN} = 0.99$). Draw a line between
    this point and the mix point, extending it to the extract
    curve. The intersection is the overall extract composi-
    tion (E_1).
4. Draw a line between F and E_1 and between R_N and S.
    It should be evident that they will intersect to the left
    and below the phase diagram. Using a long straight
    edge, extend both lines to their intersection. That is the
    difference point.
5. Note: In order to do this, tape several blank sheets
    of paper to the left hand side of the phase diagram.
    It is not necessary for these sheets to have scales
    (like Figure 19.4) unless you want to double check the
    coordinates of the difference point.
6. Now you can "step off the stages" although this will
    not look very much like a staircase! Draw a line from

E_1 along an interpolated tie line. Mark the intersection with the raffinate curve as R_1, whose composition is in equilibrium with E_1.

7. Draw another line from R_1 to the difference point. Mark the intersection with the extract curve as E_2.

8. Continue this process until you reach a raffinate composition greater than 0.99 mol W/mol. You should find that, again, it takes three stages and slightly overshoots the water purity. Record all stream compositions on your PFD.

Compare your graphical construction with Figures 19.6–19.8.

d. Explore the problem.

Solve this problem with the analytical solution (Excel "Solvent Extraction/3-Stage") and with the guided graphical solution (Excel "Solvent Extraction Graphical"). Find the amount of solvent required to obtain different levels of water purity with three stages.

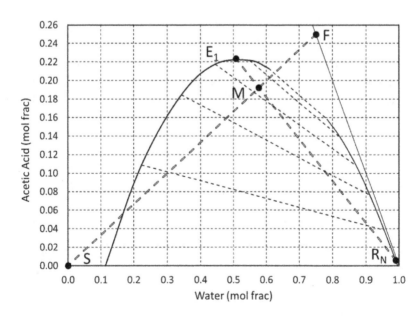

FIGURE 19.6
Phase diagram for Try This at Home 19.1 with the top and bottom streams identified. The mix point is common to the feed and product streams.

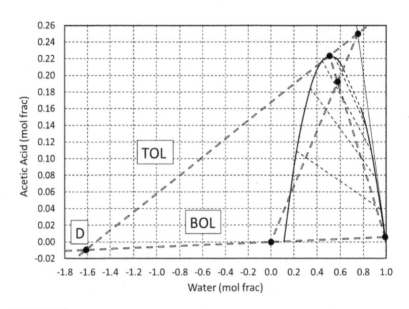

FIGURE 19.7
Phase diagram for Try This at Home 19.1 with the top and BOLs identified and the difference point marked.

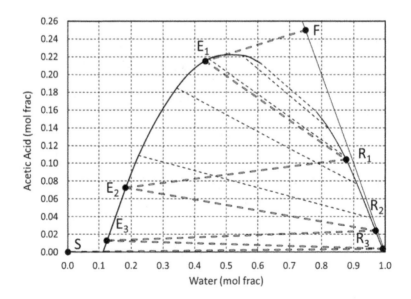

FIGURE 19.8
Phase diagram for Try This at Home 19.1 with alternating equilibrium and operating lines indicated. The difference point is not shown for clarity.

e. Be awesome!

Reconstruct the analytical and graphical solutions for the water–acetic acid–methyl tert-butyl ether system using the feed, solvent, and raffinate amounts and compositions as in this example.

Find pictures and schematics of solvent extraction processes: mixer-settlers, pulse columns, etc. and add to your collection of digital pictures.

19.2 Summary

The algebraic basis for the Hunter-Nash method observed in Chapter 18 was developed here. The purely graphical solution for a solvent extraction cascade was demonstrated. The passing stream compositions at the top and bottom of a column can be determined by an overall mole balance and knowledge that the extract and raffinate products must lie on their respective curves. A "difference point" was recognized as the intersection of lines connecting these passing streams and extending beyond the diagram. This can occur on either side. All interior passing streams connecting lines must also intersect this point. Stream compositions are found by a series of passing and leaving stream constructions on the phase diagram.

Part V

Membranes

Part V has only Chapter 20 "Membrane Separations." Separation using membranes is an important process in its own right and likely to become increasingly important as an energy-efficient process to produce potable water, treat breathing problems, purify drugs, and generate clean electrochemical energy.

But in particular, I want to show you a process that is not based on phase equilibrium. Rather, membranes separate chemicals based on their rate of transport through the membrane. I focus on two example processes, gas permeation, and reverse osmosis.

Oxygen diffuses through a gas permeation membrane faster than nitrogen. The oxygen-enriched air on the membrane's back side (permeate) can be used to treat COPD patients. The nitrogen-enriched gas on the feed side (retentate) can be used to create an inert atmosphere to control fire hazards.

Reverse osmosis drives water through a membrane faster than salt so that it can be used to make potable water. It requires large pressures to overcome osmotic pressure that makes the water want to diffuse in the opposite direction to dilute the salt water. Much of the world's population lives near the ocean in areas with little fresh water and lots of sunshine for solar energy. This is a perfect environment for desalination.

It turns out that analysis and design of membrane processes is similar to equilibrium processes, except we use the transport (flux) equations instead of phase equilibrium relationships.

20

Membrane Separations

Another class of separation processes is those using membranes. Unlike the other processes we have studied (distillation, absorption, stripping, and solvent extraction) that can be analyzed as equilibrium stages, membranes separate chemicals based on the differences in their rates of transport through the membrane. There are many types of membrane separations. We will look at two processes, gas permeation and reverse osmosis (RO).

20.1 Gas Permeation

Gas permeation is used to separate gas mixtures by passing a fraction through a membrane, driven by the difference in partial pressures of each component. Common applications include removal of acid gases (carbon dioxide and hydrogen sulfide) from natural gas (mostly methane), recovery of hydrogen from petroleum cracking streams, and separation of air to produce nitrogen as an inert blanket for hydrocarbon storage.

Let's use separation of air into oxygen and nitrogen as an example. Although this is usually accomplished by cryogenic distillation or pressure swing absorption for large-scale operations, we are tasked to design a small, portable unit for patients with breathing difficulties that require higher oxygen concentrations than the 21% found in air. High-concentration oxygen, up to 60%, can be used in conditions like pneumonia. Low-concentration oxygen (24%–28%) is used in patients with conditions like chronic obstructive pulmonary disease (COPD).

Figure 20.1 is a simple schematic diagram of this device showing values to be used in Example 20.1. The feed, F_F (mol/s), is air with a composition, y_F, of 0.21 (mol O_2/mol) and the balance nitrogen (we will ignore the other minor components of air). Since we are considering only two components, it is sufficient to label the mole fraction of only one – they add up to 1.0. The chamber is separated by a thin polymer membrane. Both oxygen and nitrogen diffuse though the membrane, but oxygen diffuses faster. This process (gas permeation) results in a stream of permeate, F_P (mol/s), enriched in oxygen, y_P (mol O_2/mol), and a stream of retentate, F_R (mol/s), depleted of oxygen, y_R (mol O_2/mol). The pressure on the feed/retentate side of the membrane, p_R (cm Hg), is higher than the permeate side, p_P (cm Hg). This same

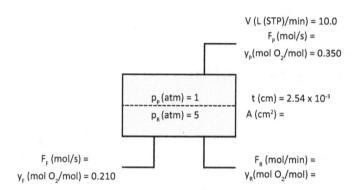

FIGURE 20.1
Labeled process flow diagram of the laboratory oxygen concentrator described in Example 20.1.

process is more commonly used to make a relatively pure nitrogen retentate (95–99.5 mol%) for inert atmospheres.

The flow rate of gas (F_P) per unit area (A) through the membrane is called the flux, J ($mol/s/cm^2$). It can be written in terms of the individual species (let's generalize for the moment as chemicals "A" and "B") as

$$J_A = F_P y_P / A \tag{20.1}$$

$$J_B = F_P (1 - y_P) / A \tag{20.2}$$

Flux is proportional to the driving force divided by the thickness, t, of the membrane. In gas permeation, the driving force is the concentration difference, expressed as partial pressure ($p_P y_P$ and $p_R y_R$ for O_2), between the permeate and the retentate. The proportionality constant is the permeability (P_A, P_B) that is a physical property of the membrane and has units (cm^3 (STP) $cm/cm^2 s$ cm Hg).

$$J_A = \frac{P_A}{t} (p_R y_R - p_P y_P) \tag{20.3}$$

$$J_B = \frac{P_B}{t} \left[p_R (1 - y_R) - p_P (1 - y_P) \right] \tag{20.4}$$

The flux equations are then

$$J_A = F_P y_P / A = \frac{P_A}{t} (p_R y_R - p_P y_P) \tag{20.5}$$

$$J_B = F_P (1 - y_P) / A = \frac{P_B}{t} \left[p_R (1 - y_R) - p_P (1 - y_P) \right] \tag{20.6}$$

In analyzing a gas permeation process, the flux equations will replace the phase equilibrium equations we used for equilibrium stage separation processes.

The permeability of various gases through several types of membranes can be found in the *Membrane Handbook* (1992). The units (cm³ (STP) cm/cm²s cm Hg) require careful consideration. Inserting them into the flux equation (repeated here for clarity) along with unit conversions, we find

$$J_A \quad = \quad F_P y_P / A$$

$$\frac{\text{mol A}}{\text{s cm}^2} \quad [=] \quad \frac{\cancel{\text{mol}}}{\text{s}} \left| \frac{\text{mol A}}{\cancel{\text{mol}}} \right| \frac{1}{\text{cm}^2}$$

$$= \quad (P_A/t)(p_R y_R - p_P y_P)$$

$$[=] \quad \frac{\cancel{\text{cm}^3\,(\text{STP})\,\text{cm}}}{\text{cm}^2\,\text{s}\,\cancel{\text{cm Hg}}} \left| \cancel{\text{cm Hg}} \right| \frac{1}{\cancel{\text{cm}}} \left| \frac{\text{mol A}}{\cancel{22.4\,\text{L (STP)}}} \right| \frac{\text{L}}{\cancel{10^3\,\text{cm}^3}}$$

Since these units are somewhat awkward and the values so small, the permeability is often expressed in Barrer (10^{-10} cm³ (STP) cm/cm²s cm Hg).

EXAMPLE 20.1   Oxygen Concentrator Using Gas Permeation

Air at 5.0 atm and 25°C is fed to a laboratory membrane oxygen concentrator. The membrane is a new material, "Oxymore Membrane," developed by BAMAChem. It will be used in a portable medical device to produce oxygen-enriched air. The thickness is 1.0 mil (2.54×10^{-3} cm) and the permeability of oxygen is 100 Barrer with a selectivity of oxygen, $\alpha_{O_2/N_2} = \dfrac{P_{O_2}}{P_{N_2}}$, of 5 over nitrogen. We need to produce 10 L (STP)/min of enriched air with 35% oxygen at 1.0 atm. What flow rates and compositions do you expect? What area of membrane is required?

If the gas is depleted of oxygen as it passes along the membrane, we might expect the concentration to depend on position. Also, a concentration gradient could exist between the membrane surface and the bulk gas. These are complications we will not consider. Assume the gas is perfectly mixed on both sides of the membrane.

a. Draw a process flow diagram.

   The process flow diagram (PFD) was shown in Figure 20.1. All known variables are labeled.

b. Perform a degree of freedom analysis.

   Unknowns (5): F_F, F_P, F_R, y_R, A

   Equations (5):

   2 mole balances (total, O_2)
   2 flux equations (O_2, N_2)
   ideal gas law (to calculate F_P).

   There are zero degrees of freedom (DOF) and the problem can be solved.

c. Write the equations and solve.

   Total MB: $F_F = F_P + F_R$

   O_2 MB: $F_F y_F = F_P y_P + F_R y_R$

   O_2 flux: $F_P y_P / A = (P_{O_2}/t)(p_R y_R - p_P y_P)$

   N_2 flux: $F_P(1 - y_P)/A = (P_{N_2}/t)[p_R(1 - y_R) - p_P(1 - y_P)]$

   Ideal gas law: $pv = F_P RT$

   In these equations, we already know v = 10 L (STP)/min, $y_F = 0.21$ mol O_2/mol, $y_P = 0.35$ mol O_2/mol, $P_{O_2} = 100$ Barrers, $P_{N_2} = 20$ Barrers, $t = 2.54 \times 10^{-3}$ cm, $p_R = 5$ atm, and $p_P = 1$ atm.

   Note how similar this solution is to equilibrium-based separation processes. All we did was replace the phase equilibrium equations with the flux equations. These equations are solved in Excel "Membrane/Ox Conc." The flow rates and compositions are $F_F = 24.2$ mmol/s, $F_P = 7.4$ mmol/s, $F_R = 16.7$ mmol/s, $y_R = 0.148$ mol O_2/mol, and A = 50.1 m². Add these values to Figure 20.1 and verify the above equations by hand calculations. Don't forget unit conversions!

d. Explore the problem.

   Systematically vary the parameters in Excel "Membrane/Ox Conc" such as pressure, membrane area/thickness, and target permeate composition. Prepare graphs of performance and discuss the results with your study partners.

e. Be awesome!

   Design the actual unit as a portable device to provide oxygen-enriched air to COPD patients. How would you pressurize and pretreat the feed air? How would you pack so much membrane area in a portable device (hint: look up hollow fiber membranes)? How could you support the thin membrane from rupturing under the pressure differential (hint: look up composite membranes)? How would you power and control the unit? What could go wrong? What are some of the competing technologies? Make a professional presentation to a mock panel of potential investors. Add pictures of membrane units to your digital image collection.

TRY THIS AT HOME 20.1   Production of Nitrogen for an Inert Gas Blanket

The atmosphere in a 1,000 gal hydrocarbon storage tank must be filled with inert nitrogen (no more than 5% oxygen) to prevent a combustible vapor. Normally, the flow of nitrogen can be small to maintain a low-oxygen concentration. However, when discharging the tank, air could be sucked into the tank, creating an extremely hazardous situation. The nitrogen flow must be large enough to make up for the volume of displaced hydrocarbon. This tank can be emptied in one hour, requiring at least a nitrogen volumetric flow of 1 L/s.

Use the same assumptions as in Example 20.1. Design a separator using BAMAChem's Oxymore membrane that can generate 1 L/s of 95% nitrogen at 2 atm using air at 10 atm. What are the expected flow rates and compositions? How much membrane area is required?

a. Draw a labeled process flow diagram.
   This has been started in Figure 20.2. Fill in values for all known variables and leave the rest blank for now.

b. Conduct a degree of freedom analysis.

Unknowns (5):

Equations (5):

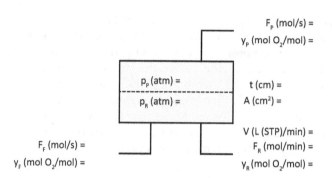

FIGURE 20.2
Labeled process flow diagram of an inert gas generator for use in Try This at Home 20.1. Fill in all known values and leave the rest blank for now.

You should find 5 unknowns and 5 equations. The problem is solvable.
c. Write the equations and solve.
I'll help by identifying these equations. Write them here.

Overall mole balance:

Oxygen mole balance:

Oxygen flux:

Nitrogen flux:

Ideal gas law:

In these equations, we already know $v = 1\,L$ (STP)/s, $y_F = 0.21\,mol\ O_2/mol$, $y_R = 0.05\,mol\ O_2/mol$ (no more than this), $P_{O_2} = 100$ Barrers, $P_{N_2} = 20$ Barrers, $t = 2.54 \times 10^{-3}\,cm$, $P_R = 10$ atm, and $p_P = 2$ atm. Insert these values into Excel "Membrane/Ox Conc" and run Solver. What the heck? It will say that the problem can't be solved! Could it be that the equations are written incorrectly? Or could it be that we are asking this membrane to do something physically impossible? Let's explore this just a bit.
d. Explore the problem.
I hope you remember that a single distillation stage (flash unit) had limits on vapor and liquid compositions that could be obtained. The most concentrated liquid occurred when taking only a drop of liquid product. The most

concentrated vapor occurred with only a bubble of vapor product. We could see this graphically using an equilibrium xy diagram and an operating (mole balance) line.

A similar situation occurs with a membrane separator except the limitation occurs because of the flux equations. An O_2 mole balance yields

$$F_F y_F = F_P y_P + F_R y_R \qquad (20.7)$$

and solving for y_R the balance becomes

$$y_R = -\frac{F_P}{F_R} y_P + \frac{F_F}{F_R} y_F \qquad (20.8)$$

If the cut is defined as the ratio of the permeate to the feed

$$\theta \equiv \frac{F_P}{F_F} \qquad (20.9)$$

then a little bit of algebra yields a design equation (see if you can work this out!)

$$y_R = \frac{-\theta}{1-\theta} y_P + \frac{y_F}{1-\theta} \qquad (20.10)$$

This equation gives all mathematically possible pairs of retentate and permeate O_2 mole fractions, some physically possible and some not (for instance, negative mole fractions). If the feed mole fraction and cut are considered constant, this will be a straight line with a negative slope that can be plotted on a y_R versus y_P graph like Figure 20.3. It is independent of membrane properties and flow rates. The design equation for any value of cut passes through (y_F, y_F).

Solving the two flux equations together (two equations/ two unknowns) will yield pairs of retentate and permeate O_2 mole fractions that satisfy both equations. Dividing the O_2 flux equation (Eq. 20.5) by the N_2 flux equation (Eq. 20.6) yields

$$\frac{y_P}{1-y_P} = \frac{P_{O_2}}{P_{N_2}} \left(p_R y_R - p_P y_P\right) \Big/ \left[p_R\left(1-y_R\right) - p_P\left(1-y_P\right)\right] \qquad (20.11)$$

Notice that this equation is independent of flow rates and membrane size because these variables cancelled out.

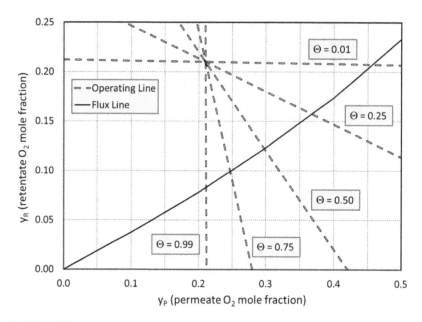

FIGURE 20.3
Graphical solution for a gas permeation process as described in Try This at Home 20.1 with various cuts. Notice that all operating lines pass through the feed composition $(yF, yF) = (0.21, 0.21)$ with a slope of $-\Theta/(1 - \Theta)$.

The equation depends only on the membrane properties (permeability) and the process pressures. This can also be plotted in Figure 20.3. Where the two lines intersect, the pairs (y_P, y_R) satisfy both the mole balance equations and the flux equations.

We will relax the retentate concentration as a variable and add the cut as a constraint. If we set the cut to 0.99, only 1% of the feed is left in the retentate and it contains the lowest possible O_2 concentration, 8.3 mol%. Ah, so that is why we couldn't reach 5 mol%! This relatively large flow rate through the membrane will require the largest membrane area. The highest permeate O_2 concentration occurs when we pass only 1% of the feed yielding 45.9 mol% O_2. So the concentration limits are 8.3–21.0 mol% O_2 in the retentate and 21.0–45.9 mol% O_2 in the permeate. Vary the cut from 1% to 99% using Excel "Membrane/Ox Conc" and plot product mole fraction. You can also calculate area and flow rates and plot them against cut.

e. Be awesome!

Some separations require multiple stages in series. For example, to make fissile uranium for the first atomic bombs, $^{235}UF_6$ was separated from $^{238}UF_6$ using membranes. The rate of diffusion is largely determined by the compound molecular weight. With about a 1% difference in weight between the isotopes the selectivity is only 0.4%. So multiple stages (actually thousands!) with interstage compression and cooling were required and housed in some of the world's biggest buildings. Other separation techniques have largely replaced this very expensive process. It is a fascinating story that you should read.

So here is a challenge that I have confidence you are awesome enough to handle. Add an identical second membrane separator to further refine the O_2-depleted retentate product in this example using Excel "Membrane/2-Stage DIY." Input data and equations, letting the cut be a variable and fixing $y_{R2} = 0.05$. You should find that 558 L (STP)/min of air and a cut of 67.2% (both membrane units) is required to produce 60 L (STP)/min of N_2 containing 5 mol% O_2. The membrane areas are 1,032 and 383 m². The full solution is presented in Excel "Membrane/2-Stage." The PFD for this membrane process is shown in Figure 20.4.

If the permeate stream was being refined, compression (and perhaps cooling) between separators would be required because the low-pressure permeate from the first separator becomes the high-pressure feed to the second separator.

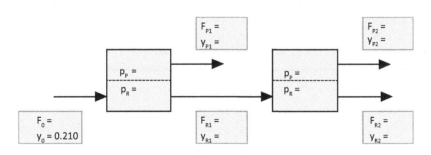

FIGURE 20.4
PFD for two-stage N_2 generator.

20.2 Reverse Osmosis

Another important membrane separation process is RO and a major application is water treatment, for example, desalination of sea water. RO processes
use pressure to drive solvents through the membrane, leaving dissolved
solutes behind in the retentate. If pure water and salt water are placed on
opposite sides of a semipermeable membrane, the pure water will diffuse
through the membrane to dilute the salt water – driven by osmotic pressure, π.
That process is called osmosis and the water is going in the opposite direction than we want for desalination. To reverse the water flow we must apply
a pressure on the retentate side greater than the osmotic pressure, hence the
name reverse osmosis.

The flux of water to the permeate side, J_W, is usually expressed in volumetric units (L/m²/h)

$$J_W = \frac{F_P'(1-x_P)}{A\rho} \tag{20.12}$$

where F_P' (kg/h) is the mass flow rate of permeate (low salt water), x_P (kg S/kg)
is the mass fraction of salt in the permeate, A (m²) is the membrane area, and
ρ (kg/L) is the density of the water. SI units (well, L and h also) are used here
but in practice units such as gal/ft²/day are often used.

The flux of salt to the permeate side, J_S, is usually written in mass units
(kg/m²/h).

$$J_s = \frac{F_P' x_P}{A} \tag{20.13}$$

Flux is proportional to the driving force divided by the thickness, t, of
the membrane. In RO, the driving force for the solvent – water in this
case – is the difference between the pressure drop across the membrane
$(p_R - p_P)$ and the osmotic pressure difference $(\pi_R - \pi_P)$ between the retentate and permeate ($\pi_P = 0$ if there is essentially no salt in the permeate). The
proportionality constant is the permeability (P_W, P_S), which is a physical
property of the membrane. The osmotic pressure is a physical property of
the solution.

$$J_W = \frac{P_W}{t}(\Delta p - \Delta \pi) = \frac{P_W}{t}\left[(p_R - p_P) - (\pi_R - \pi_P)\right] \tag{20.14}$$

The unit of Δp and $\Delta \pi$ will usually be bar or atm (almost the same value). We
will assume the osmotic pressure is a linear function (with slope "a") of the
salt concentration expressed as mass fraction. This is more likely valid at low
concentrations.

$$\pi_R = ax_R, \pi_P = ax_P, a[=]\text{bar} \tag{20.15}$$

The driving force for the salt is the concentration difference across the membrane expressed as mass concentration (kg/L).

$$J_S = \frac{P_S}{t}\Delta x = \frac{P_S}{t}(c_R - c_P) \tag{20.16}$$

You will often encounter permeance, K_W and K_S, which is the permeability divided by the thickness of the active layer – a more fundamental membrane property. The flux equations are

$$J_W = \frac{F'_P(1-x_P)}{A\rho} = K_W\left[(p_R - p_P)-(\pi_R - \pi_P)\right] \tag{20.17}$$

$$J_s = \frac{F'_P x_P}{A} = K_S(c_R - c_P) \tag{20.18}$$

Attention must be paid to the units of permeance. A dimensional analysis of Eqs. 20.12 and 20.13 shows

$$\frac{\text{Kg}}{\text{h}}\left|\frac{}{\text{m}^2}\right|\frac{\text{L}}{\text{Kg}}[=]\frac{K_W}{}\left|\text{bar}\right. \qquad K_W[=]\frac{\text{L}}{\text{m}^2 \text{ h bar}}$$

$$\frac{\text{Kg}}{\text{h m}^2}[=]\frac{K_s}{}\left|\frac{\text{kg}}{\text{L}}\right. \qquad K_s[=]\frac{\text{L}}{\text{m}^2 \text{ h}}$$

In analyzing an RO process, as with gas permeation, the flux equations will replace the phase equilibrium equations we used for equilibrium stage separation processes.

The mass fraction of salt on the retentate side of the membrane can be considered uniform if the system is well mixed. However, salt concentration can build up near the wall as water passes through the membrane, leaving the salt behind. The salt then diffuses back to the bulk retentate. This has the effect of increasing the osmotic pressure at the membrane surface and Eq. 20.12 shows that this decreases the flux of water. (This does not happen on the permeate side – can you see why?) This effect is called concentration polarization and we will account for it with a simple factor M such that

$$x_{RW} = Mx_R \tag{20.19}$$

where x_{RW} is the mass fraction of salt at the retentate membrane surface and the flux equations become

$$J_W = \frac{F_P'(1-x_P)}{A\rho} = K_W\left[(p_R - p_P) - a(Mx_R - x_P)\right] \tag{20.20}$$

$$J_S = \frac{F_P'x_P}{A} = K_S(Mc_R - c_P) \tag{20.21}$$

Correlations exist to predict this concentration polarization factor.

As the salt water passes through the membrane device, it becomes more concentrated in salt, also increasing the osmotic pressure on the retentate side and decreasing the water flux. This is strongly dependent on the device design and will not be analyzed here. We will continue to assume the bulk fluid is well mixed.

Of course the above discussion will apply to any solvent/solute pair to be separated by RO. The analysis of an RO process is very similar to that for gas permeation. The flow rates and compositions of all streams will depend on mass balances, flux equations, and various process specifications.

EXAMPLE 20.2 Desalination of salt water

A small town in southern California will purchase a 50,000 gal/day SWRO (sea water reverse osmosis) system from BAMAChem to provide drinking water. The sea water has a TDS (total dissolved solids) content of 3.5 wt% (assume all NaCl). The sea water is pumped to 1,000 psi before entering the SWRO and delivers fresh water at 50 psi. The salt rejection (percentage staying in the retentate) is 99.5% with a cut (water recovery) of 50%. The water permeance is 2.0 $L/m^2/h/bar$ and the salt permeance is 0.05 $L/m^2/h$. If each of the ten membrane cartridges can produce 5,000 gal/day, what is the active membrane surface area in a cartridge? Determine all flow rates and compositions.

There is no mention of concentration polarization in this problem. So we will assume perfect mixing on the retentate side and bulk concentration at the membrane surface.

a. Draw a labeled process flow diagram.
   The PFD is shown in Figure 20.5. Values are assigned to all known variables.

b. Conduct a degree of freedom analysis.

   Unknowns (6): F_F, F_P, x_P, F_R, x_R, A

   Note: Equations 20.15 and 20.16 show that we also need π_P, π_R, c_P, c_R, and (to determine concentrations) ρ_P and ρ_R.

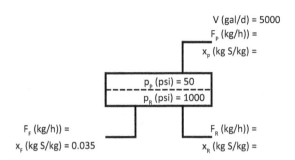

V (gal/d) = 5000
F_P (kg/h)) =
x_P (kg S/kg) =

p_P (psi) = 50
p_R (psi) = 1000

F_F (kg/h)) =
x_F (kg S/kg) = 0.035

F_R (kg/h)) =
x_R (kg S/kg) =

FIGURE 20.5
PFD of the salt water RO process described in Example 20.2. Known values are inserted and unknown values left blank for now.

We could list these as unknowns or just note that we have osmotic pressure and density relationships (from Perry's Handbook) to calculate these from x_R and x_P, leaving them as the more fundamental unknowns (see below).

Equations (7):

2 mass balances (total and salt)

2 flux equations (water and salt)

2 process specifications (cut and rejection)

1 process physical property (density, to calculate Fp (kg/h) from 5,000 gal/d)

That is 7 equations (or relationships) for 6 unknowns. The problem is overspecified. Good thing we checked the DOF! The mass balances and flux equations are non-negotiable and the 5,000 gal/d was our basis. Looks like we can't specify both cut and salt rejection. Let's drop the salt rejection specification and just calculate salt rejection from the results.

c. Write the equations and solve.

Total mass balance: $F_F = F_P + F_R$

Salt mass balance: $F_F(0.035) = F_P x_P + F_R x_R$

Water flux: $F_P(1 - x_P)/(A\rho) = K_W[(p_R - p_P) - (\pi_R - \pi_P)]$

Salt flux: $F_P x_P/A = K_S(c_R - c_P)$

Process Specification (cut): $F_P = 0.5 F_F$

Additional equations:

$\pi(atm) = 45.61x' + 0.01$, where x' [=] mol S/mol Physical property

$x' = (x/58.44)/(x/58.44 + (1 - x)/18.02)$ (changing from mass to mole basis)

$\rho(kg/L) = 0.7143x + 1.00$ (physical property)

$c(kg\ S/L) = x\rho$

These equations are included in Excel "Membrane/SWRO." We already know that $v = 5{,}000$ gal/day, $x_F = 0.035$ mol S/mol, $p_R = 1{,}000$ psi, $p_P = 50$ psi, $K_W = 2.0$ L/m²/h/bar, $K_S = 0.05$ L/m²/h, and cut = 50%. After inputting these values, the solution will be $F_F = 1{,}577$ kg/h, $F_P = 789$ kg/h, $x_P = 2.89 \times 10^{-5}$ kg S/kg, $F_R = 789$ kg/h, $x_R = 7.09 \times 10^{-2}$ kg S/kg, and $A = 6.12\,m^2$.

d. Explore the problem.

Drinking water standards limit the TDS to 500 ppm. Translate ppm to weight fraction of salt in the permeate. Revise Solver to set the outlet concentration and relax the percent cut. Explore the effect of other variables on the SWRO performance.

e. Be awesome!

Copy Excel "Membrane/SWRO" to another page and add a second cartridge in series. The total area of the two cartridges should be the same as the single cartridge. Don't forget that permeate from the first cartridge will need to be pumped back up to the original pressure. Compare the performances.

20.3 Summary

Membranes separate chemical mixtures by the differences in rates of transport. There are many types of membrane processes, but we just analyzed two: gas permeation and reverse osmosis. The analysis is surprisingly similar to the equilibrium stage processes that we studied, except that flux equations are used instead of phase equilibrium equations. Different units and properties are used in industrial practice, but generally flux (amount transferred per area per time) is proportional to a composition or pressure-driving force. The flux equations, mole or mass balances, and various process specifications were solved simultaneously using Excel's Solver tool.

Appendix: Properties of Chemicals Used in This Book (Adapted from Felder et al., 2015)

Physical Properties

Name	Formula	MW (g/mol)	Tb(°C)		SG (20°/4°)
Acetic acid	CH_3COOH	60.05	118.2	24.39	1.049
Acetone	C_3H_6	58.08	56	30.2	0.791
Benzene	C_6H_6	78.11	80.1	30.77	0.879
Butane	C_4H_{10}	58.12	−0.6	22.31	
Carbon dioxide	CO_2	44.01	Sublimes at −78°C		
Chloroform	$CHCl_3$	119.39	61	—	1.489
Ethanol	C_2H_5OH	46.07	78.5	38.58	0.789
Ethyl benzene	C_8H_{10}	106.16	136.2	35.98	0.867
Hexane	C_6H_{14}	86.17	68.7	28.85	0.659
Isopropyl acetate	$C_5H_{10}O_2$	102.13	89	32.1	0.87
Methyl tert-butyl ether	$C_5H_{12}O_2$	88.15	55.2	27.9	0.7404
Nitrogen	N_2	28.02			
Octane	C_8H_{18}	114.22	125.5	—	0.703
Oxygen	O_2	32			
o-Xylene	C_8H_{10}	106.16	144.4	36.82	0.88
Sodium chloride	NaCl	58.45			2.163
Toluene	C_7H_8	92.13	110.6	33.47	0.866
Water	H_2O	18.016	100.0	40.66	1.00

Heat Capacities at Constant Pressure

				Cp(kJ/mol °C)=a+bT+cT²+dT³				
	Liquid			Gas				
Name	$a \times 10^3$	$b \times 10^5$	Range (°C)	$a \times 10^3$	$b \times 10^5$	$c \times 10^8$	$d \times 10^{12}$	Range (°C)
Acetone	123.0	18.6	−30 to 60	71.96	20.10	−12.78	34.76	0–1200
Benzene	126.5	23.4	6–67	74.06	32.95	−25.20	77.57	0–1200
Butane				92.3	27.88	−15.47	34.98	0–1200
Carbon dioxide				36.11	4.233	−2.287	7.464	0–1500
Ethanol	103.1	55.7	0–100	61.34	15.72	−8.749	19.83	0–1200
Hexane	216.3		20–100	137.44	40.85	−23.92	57.66	0–1200
Nitrogen				29.00	0.2199	0.5723	−2.871	0–1500
Oxygen				29.10	1.158	−0.6076	1.311	0–1500
Toluene	148.8	32.4	0–110	94.18	38.00	−27.86	80.33	0–1200
War	74.4		0–100	33.46	0.6680	0.7604	−3.593	0–1500

		Antoine's Constants			
		$\log_{10}P^* = A - B/(T+C)$ P [=] mmHg, T [=] °C			
Name	A	B	C	D	Range
Acetic acid	7.38782	1533.313	222.309		29.8–126.5
Acetone	7.11714	1210.595	229.664		−12.9 to 55.3
Benzene	6.89272	1203.531	219.888		14.5–80.9
Butane	6.82485	943.453	239.711		−78.0 to −0.3
Chloroform	6.90328	1161.03	227.4		−30 to 150
Ethanol	8.11220	1592.864	226.184		19.6–93.4
Ethyl benzene	6.95650	1423.543	213.091		56.5–137.1
Hexane	6.88555	1175.817	224.867		110–69.5
Octane	6.91874	1351.756	209.100		52.9–126.6
o-Xylene	7.002	1476.393	213.872		615–145.4
Toluene	6.95805	1346.773	219.693		353–111.5
Water	7.96681	1668.210	228.000		60–150
		Henry's Law Constants			
		H = exp (A/T + B 1nT + CT + D) H [=] psi, T [=] K			
Carbon dioxide	−1392	0.1210	0.01479	3.984	

References

Basmadjian, D., *Mass Transfer and Separation Processes*, 2nd ed., CRC Press, Boca Raton, FL, 2007.

Benitez, J., *Principles and Modern Applications of Mass Transfer Operations*, 3rd ed., Wiley, New York, 2016.

Duss, M., and Taylor, R., "Predict Distillation Tray Efficiency," Chem. Eng. Prog., **114** (7), 24–30 (2018).

Eckert J. S., "Selecting Proper Distillation Column Packing," Chem. Eng. Prog., **66** (3), 39 (1970).

Fair, J. R., and R. L. Matthews, "Better Estimate of Entrainment from Bubble Caps," Pet. Refiner, **37** (4), 153 (1958).

Felder, R. M., R. W. Rousseau, and L. G. Bullard, *Elementary Principles of Chemical Engineering*, 4th ed., Wiley, New York, 2015.

Geankoplis, C. J., A. A. Hersel, and D. H. Lepek, *Transport Processes and Separation Process Principles*, 5th ed., Prentice Hall, Englewood Cliffs, NJ, 2018.

Green, D.W. and M. Z. Southard, Perry's Chemical Engineers Handbook, 9th ed., McGraw-Hill, New York, 2019.

Hunter, T. G., and A. W. Nash, "The Application of Physico-Chemical Principles to the Design of Liquid-Liquid Contact Equipment, Part II: Application of Phase-Rule Graphical Method," *J. Soc. Chem. Ind.*, **53**, 95T–102T (1934).

King, C. J., *Separation Processes*, 2nd ed., Dover Publications, New York, 2013.

Kister, H. Z., *Distillation Design*, McGraw-Hill, New York, 1992.

Kister, H. Z. and D. R. Gill, "Predict Flood Points and Pressure Drop for Modern Random Packings," Chem. Eng. Prog., **87** (2), 32–42 (1991).

Kister, H. Z., J. Scherffius, K. Afshar, and E. Abkar, "Realistically Predict Capacity and Pressure Drop for Packed Columns," Chem. Eng. Prog., **103** (7), 28–38 (2007).

Lane, A. M., "Celebrating ChE in Song," Chem. Eng. Ed., **42**, 52 (2008).

Leva, M., "Reconsider Packed-Tower Pressure-Drop Correlations," Chem. Eng. Prog., **88** (1), 65–72 (1992).

Lewis, W. K., "The Efficiency and Design of Rectifying Columns for Binary Mixtures," Ind. Eng. Chem., **14**, 492 (1922).

McCabe, W. L., J. C. Smith, and P. Harriott, *Unit Operations of Chemical Engineering*, 7th ed., McGraw-Hill, New York, 2004.

McCabe, W. L., and E. W. Thiele, "Graphical Design of Fractionating Columns," Ind. Eng. Chem., **17**, 602 (1925).

Miao, X., H. Zhang, T. Wang, and M. He, "Liquid–Liquid Equilibria of the Ternary System Water+Acetic Acid+Methyl *tert*-Butyl Ether," J. Chem. Eng. Data, **52** (3), 789–793 (2007).

Murphree, E. V., "Graphical Rectifying Column Calculations," Ind. Eng. Chem., **17**, 960 (1925).

O'Connell, H. E., "Plate Efficiency of Fractionating Columns and Absorbers," Trans. AIChE, **42**, 741–775 (1946).

Peters, W. A. Jr., "The Efficiency and Capacity of Fractionating Columns," Ind. Eng. Chem., **14** (6), 476–479 (1922).

Seader, J. D., E. J. Henley, and D. K. Roper, *Separation Process Principles*, 4th ed., Wiley, New York, 2015.

Sherwood, T. K., G. H. Shipley, and F. A. L. Holloway, "Flooding Velocities in Packed Columns," Ind. Eng. Chem., **30** (7), 765–769 (1938).

Sørensen, J. M. and Arlt, W., *Liquid-Liquid Equilibrium Data Collection*. DECHEMA, Deutsche Gesellschaft für Chemisches Apparatewesen, Great Neck, N.Y., distributed exclusively by Scholium International, Frankfurt/Main, 1979.

Souders, M. and G. G. Brown, "Design of Fractionating Columns, Entrainment and Capacity," Ind. Eng. Chem., **38** (1), 98–103 (1934).

Strigle, R. F. Jr., *Packed Tower Design and Applications*, 2nd ed., Gulf Publishing, Houston, TX, 1994.

Walas, S. M., *Phase Equilibria in Chemical Engineering*, Butterworth Publishers, London, 1985.

Wankat, P. C., *Separation Process Engineering*, 4th ed., Prentice Hall, Englewood Cliffs, NJ, 2016.

Epilogue

So, how do you feel right now? A little "separation anxiety" perhaps? You now understand how separation processes work and how to make a preliminary design. You also have an appreciation for just how much you don't know. A little knowledge is only dangerous when you think you know everything. My favorite chemical reaction engineering author, Professor Scott Fogler, quoted Winston Churchill, saying, "This is not the end. It is not even the beginning of the end. But it is the end of the beginning."

I started writing the book in the summer of 2016, right after retiring, and while recovering from a serious illness. Confined to a wheelchair and later working in-between relentless physical therapy sessions, I began by re-reading the many books on separation processes in my personal library. Drawing on 30 years of teaching separations, I took notes on the essential concepts that I knew you needed and could master in a one-semester course. I then wrote the lessons exactly the way I would teach you in person, with lots of worked and guided examples, and keeping K.I.S.S. principles in mind.

If you are going to be a professional chemical engineer, I encourage you to never stop learning. While writing this textbook I learned several new things, even though I have taught this course longer than you've been alive! I didn't know the work of Duss and Taylor on efficiency and the O'Connell correlation until my email discussions with Dr. Taylor. I thought I had invented Figure 20.3 showing the limits of membrane separation until I came upon the same development in a later edition of Dr. Phillip Wankat's book. I never realized the difficulty setting up the Lewis method for distillation of non-ideal solutions in Excel without circular references. Until now, I had never actually tried to solve the NRTL equations. As it turns out, you can teach an old dog new tricks!

Something I intentionally left out of this book is mass transfer-based methods. These can be used for estimating the number of equilibrium stages, tray efficiencies, packing HETPs, etc. This would be a fundamentally better approach, especially for packed columns that have no discrete stages. What I learned from several discussions and many papers is that these methods are standard for a few specific applications, but for most situations they just aren't as convenient and accurate as the equilibrium stage methods and empirical correlations. Maybe they'll show up in a future edition of this book!

I hope that I demystified the design and analysis of separation processes for you. All of the processes that can be treated as equilibrium stages are described by just a few equations: material and energy balances, phase equilibrium relationships, and miscellaneous process specifications. You can apply this knowledge to analyze virtually any variation of the common separation processes to answer the fundamental question, "how many equilibrium stages are required to achieve a degree of separation?" Then, using experimental data, empirical correlations, and rules of thumb (heuristics) you can decide how many actual trays or height of packing is needed along with the necessary column diameter. This analysis is easily done with commercial chemical process simulators, but now you know what they are doing behind that computer interface and so are better equipped to interpret their results and understand their limitations.

If you would like to go beyond "Separation Process Essentials," many other textbooks are more comprehensive and detailed. Most of the techniques I described were developed in the early 20th century and have been active areas of research and development ever since. You can imagine the literature on separation processes is immense and just waiting for you to dive in.

Please make use of the course website *SeparationsBook.com* and feel free to contribute by participating in the discussion boards or sending me suggestions to improve the book, interesting problems to work, and to simply correct any mistakes you find (and will no doubt be legion!)

I am indebted to Professor Phillip Wankat (*Separation Process Engineering*) and Professors J. D. Seader, Ernest J. Henley, and D. Keith Roper (*Separation Process Principles*) who wrote the two excellent textbooks that I am most familiar with, having used them in my course for decades. Henry Z. Kister, senior fellow and director of fractionation technology at Fluor Corporation, kindly reviewed Chapter 16 on column design and made many good suggestions. My colleagues at The University of Alabama, particularly Professors Steve Ritchie, Tonya Klein, David Arnold, Marvin McKinley, Jason Bara, and Heath Turner, have taught me much about separations over the years and engaged with me in debating how to present certain material. I am especially thankful to Professor Steven Weinman, our current separations professor, who has used the book as it developed for the past two semesters and made a very thorough review with many corrections and suggestions. Finally, I most appreciate the support of my wife, Lorrie, sons Charles, Sean, and Billy, daughter Liz, and son-in-law Chase. They have encouraged me throughout the two and a half years of writing and no doubt enjoyed my many eloquent dinner conversations about spreadsheet puzzles, thermodynamic oddities, fascinating applications ... uh, they are all staring at me with folded arms. Something tells me I better wrap this up! Bye!

Index